D1617432

THE ETHNOMETHODOLOGICAL FOUNDATIONS OF
MATHEMATICS

Studies in ethnomethodology

Edited by Harold Garfinkel
Professor of Sociology
University of California, Los Angeles

The ethnomethodological foundations of mathematics

Eric Livingston

Routledge & Kegan Paul

London, Boston and Henley

Dedicated to my parents
Herbert and Rosetta Livingston

First published in 1986
by Routledge & Kegan Paul plc

14 Leicester Square, London WC2H 7PH, England

9 Park Street, Boston, Mass. 02108, USA and

Broadway House, Newtown Road,
Henley on Thames, Oxon RG9 1EN, England

Set in Times, 10 on 11 pt
by Hope Services of Abingdon
and printed in Great Britain
by Billing & Sons Ltd., Worcester

Copyright © Eric Livingston 1986

No part of this book may be reproduced in
any form without permission from the publisher,
except for the quotation of brief passages
in criticism

Library of Congress Cataloging in Publication Data

Livingston, Eric.
The ethnomethodological foundations of mathematics.
(Studies in ethnomethodology)
Bibliography: p.
1. Mathematics – Philosophy. 2. Ethnomethodology.
I. Title. II. Series.
QA8.4.L48 1985 511.3 85-1898

British Library CIP data also available

ISBN 0-7102-0335-7

Contents

Preface ix

Acknowledgments xii

Introduction 1

A Non-Technical Introduction to Ethnomethodological Investigations of the Foundations of Mathematics through the Use of a Theorem of Euclidean Geometry 1
A Guide to the Reading of this Book 15

Part I 21

Introduction. The Phenomenon: The Existence of Classical Studies of Mathematicians' Work 23

1 A Review of the Classical Representation of Mathematicians' Work as Formal Logistic Systems 25

2 An Introduction to Gödel's Incompleteness Theorems: Their Metamathematical Interpretation Contrasted with the Proposal to Study Their Natural Accountability in and as the Lived-Work of Their Proofs 31

Part II A Descriptive Analysis of the Work of Proving Gödel's First Incompleteness Theorem 37

3 Gödel Numbering and Related Topics: Background Materials for a Proof of Gödel's Theorem 39

CONTENTS

4 The Double-Diagonalization/'Proof': Features of the Closing Argument of a Proof of Gödel's Theorem as Lived-Work   45

5 A Technical Lemma: A Lemma Used in the Proof of Gödel's Theorem; Its Origins as a Technical Residue of the Work of Proving Gödel's Theorem within that Self-Same Work   51

6 Primitive Recursive Functions and Relations: An Initial Discussion of the Irremediable Connection between a Prover's Use of the Abbreviatory Practices/Practical Techniques of Working with Primitive Recursive Functions and Relations and the Natural Accountability of a Proof of Gödel's Theorem   57

7 A Schedule of Proofs: An Extended Analysis of the Lived-Work of Producing the Body of a Proof of Gödel's Theorem   65

   A A Schedule of Proofs   65
   B A Schedule of Proofs as Lived-Work   69

      1 As it is used in developing a schedule of proofs, a Gödel numbering is not an abstractly defined correspondence between the symbols, expressions, and sequences of expressions of P and the natural numbers; it is a technique of proving   69
      2 The schedule of proofs has a 'directed' character: it leads to and is organized so as to permit, the construction of G as a primitive recursive relation   76
      3 The selection and arrangement of these-particular propositions as composing this-particular, intrinsically sequentialized order of proving is the situated achievement of the work of producing that schedule of proofs   80
         (a) Six themes concerning the lived-work of producing a schedule of proofs   81
         (b) The construction of a schedule of proofs so as to provide an apparatus within itself for the analysis of the work of its own construction   100
         (c) The correspondence between the propositions of a schedule of proofs and the syntax of formal number theory as an achievement of the schedule of proofs itself   117
         (d) A review of the work of generalizing a schedule of proofs so as to elucidate the character of the development and organization of a schedule as a radical problem, for the prover, in the production of social order   125

    4 The work of providing a consistent notation for a schedule of proofs articulates that schedule as one coherent object   137

8 A Structure of Proving   149

   A The Characterization Problem: The Problem of Specifying What Identifies a Proof of Gödel's Theorem as a Naturally Accountable Proof of Just That Theorem; The Texture of the Characterization Problem and the Constraints on Its Solution; The Characterization Problem as the *Foundational Problem*   149
   B Generalizing the Proof of Gödel's Theorem (As a Means of Gaining Technical Access to the Characterization Problem)   154
   C A Structure of Proving: The Availability to a Prover of the Proof of Gödel's Theorem as a Structure of Practices; The Proof as the Pair The-Proof/The-Practices-of-Proving-to-Which-That-Proof-is-Irremediably-Tied   169

Part III Conclusion   173

9 Summary and Directions for Further Study   175

   A Classical Studies of Mathematical Practice: A Review of the Book's Argument   175
   B Prospectus: Mathematicians' Work as Structure Building   177

Appendix   179

The Use of Ethnomethodological Investigations of Mathematicians' Work for Reformulating the Problem of the Relationship between Mathematics and Theoretical Physics as a Real-World Researchable Problem in the Production of Social Order   181

Notes   190

Bibliography   237

Preface

The early Greeks were both amazed and perplexed by mathematical proofs. On one hand, the objects of geometry were made available and described, and their properties were established, through the use of drawn figures. Yet the Greeks recognized that the geometric objects themselves had a curious, unexplicated relationship to their depiction. They were further puzzled by the fact that the mathematical proposition was demonstrated not as a matter of rhetorical argumentation, merely to the satisfaction of mathematicians immediately present. The mathematical theorem was proved as something necessarily true, a fact anonymous as to its authorship, available for endless inspection, established for all time − and this as a required feature of the actual demonstration itself. When it was proved that the 'field' of constructable lengths contained incommensurable elements, the Greeks were unable to turn away from the evidentness of the mathematical demonstration even though it went against their deepest philosophical commitments. It was said that Pythagoras, on proving this fact, committed suicide by drowning.

At the turn of this century great interest was again shown in the origins of mathematical truth and in the nature of mathematical proofs. This was stimulated, in part, by circumstances similar to the Pythagorean proof of the existence of irrational numbers: developments in set theory had led to the proved existence of a continuous, 'space-filling' curve; even more spectacularly (but received with much greater suspicion), Zermelo proved that any set could be well-ordered. As a particular case, it follows that there exists some partial ordering of the real numbers such that every nonempty subset of them has a first element. Zermelo's proof, however, did not show what that partial order is.

At the same time as these developments, elegant proofs using set-theoretic methods − such as the proof that the transcendental numbers are more numerous than the algebraic ones — were given. But it was also shown, by ingenious but elementary reasoning, that the unexplicated notion of a 'set' led to contradictions, the most famous of these being Russell's paradox. Hilbert's solution of the 'invariant problem' demonstrated the power of abstract methods but, for some, raised questions as to the sense in which the problem had actually been solved. The burgeoning development of the new geometries had invigorated philosophers' consideration of the empirical status of mathematical truth. And Hilbert's researches in geometry again showed the strength of the axiomatic method but exhibited as well flaws in the reasoning of Euclid's *Elements*.

One consequence of these origins of early twentieth-century investigations of the foundations of mathematics was regrettable. The predominating interest of those studies, at least in the received view, was to demonstrate that the methods used by practising mathematicians and the results established through ordinary mathematical practice were themselves free from criticism. The attempt was made to construct indubitable foundations for mathematical practice. In consequence, interest in the original question — what made up the evident and transcendental character of mathematical proofs? — shifted to the problem of demonstrating the incorrigibility of those same proofs.

During this time, mathematical research continued unabated, little affected by the investigations directed specifically to mathematical foundations. Moreover, it was mathematicians' daily production of ordinary, naturally accountable proofs that supplied not only the promised object but the basis of the logician's research. Although foundational studies did have an indirect effect on mathematical practice, a consequence of their most celebrated achievement, Gödel's incompleteness theorems (themselves proved in the style of ordinary mathematics), was to hasten the incorporation of mathematical logic into mainstream mathematical research, not to alter mathematical practice. The living foundations of mathematics — and, as a particular case, the origins of the adequacy of Gödel's own proofs — remained untouched and unexamined.

These remarks set in contrast the direction taken in the present work. This book is a study of the foundations of mathematics, but in the original sense. It is a study of the genetic origins of mathematical rigor, examining the proofs of ordinary mathematics and investigating how the adequacy of such proofs, for the purposes of everyday mathematical inquiry, is practically obtained. The book formulates and, in a certain sense, solves the problem of the foundations of mathematics as a problem in the local production of social order. It does this not by reviewing received philosophies of mathematics, not by proffering a theory of social action, not through an historical or cultural analysis, *but by rediscovering and exhibiting the naturally accountable mathe-*

matical proof, in its identifying detail for mathematicians, as a social achievement. This book asks and attempts to describe how mathematicians, in and as their daily work, as the *sine qua non* of their work for and in the company of mathematicians, produce the naturally accountable proofs of ordinary mathematics which, as the witnessed, local achievement of the work of their demonstration, therein exhibit both the transcendental character of that work and of the objects those proofs describe.

Eric Livingston
May 1984

Acknowledgments

This book could not have been written without the assistance, supervision and support of Harold Garfinkel. His influence has been pervasive in formulating the problem of mathematical foundations as a problem in the production of social order and in carrying out the research. His suggestions provided critical directions for further investigation, and his encouragement was steadfast.

In many different ways I am indebted to many other people. Emanuel Schegloff and Melvin Seeman provided, each in his own way, penetrating criticisms and sustaining intellectual and emotional support. Herbert Enderton and Louis Narens showed interest in my research and assisted me in learning the necessary mathematics. Conversations with Professor Enderton, and his close attention to the mathematical details of my argument, were invaluable; Professor Narens's approval of the direction of my investigations — if not, perhaps, his agreement with their results — was a source of intellectual sustenance. I thank as well Jeffery Alexander, Donald Babbitt, Robert Blattner, Phillip Bonacich, Henry A. Dye, Rod Harrison, Rudolph De Sapio, V.S. Varadarajan and Robert Westman. One of my brothers, Charles Livingston, himself a professional mathematician, gave unstintingly of his time to further my studies and was supportive of them over their entire course.

Over the course of my undergraduate and graduate studies I was fortunate in being part of a local culture of students and faculty engaged in studies of naturally organized ordinary activities. I take this occasion to express my indebtedness to them, and I thank, in particular, Melinda Baccus, Stacy Burns, Trent Eglin, Richard Fauman, Harold Garfinkel, George Girton, Gail Jefferson, Ken Liberman, Michael Lynch, Douglas MacBeth, Ken Morrison, Christopher Pack, Anita Pomerantz, Britt Robillard, Friedrich Schrecker, the late Harvey Sacks, Emanuel Schegloff, Dave Sudnow and Larry Wieder.

I thank my friends not already mentioned and, in particular, Paul

ACKNOWLEDGMENTS

Colomy and Maureen McConaghy. I thank Mary Takami, the staff of the Department of Sociology, UCLA, and the Department of Mathematics, UCLA (for unrestricted use of the Department of Mathematics' Reading Room). I thank Diane Wells, Librarian for the Department of Philosophy's Reading Room, for her assistance in locating historical materials. Special thanks go to Ralph Edwards, Ted Regler and Ralph Edwards Productions for their encouragement and unsolicited support.

Readers of this book will notice my debt, particularly in later chapters, to a book by Joel W. Robbin, *Mathematical Logic: A First Course*. I am grateful to Benjamin/Cummings Publishing Company for granting me permission to draw on it as I have. I am grateful as well to Harcourt Brace Jovanovich, Inc. for granting me permission to reproduce the illustration that appears on page 145.

An Exxon Research Fellowship in the Program in Science, Technology, and Society at M.I.T. provided support for continuing my studies of the discovering sciences. During my fellowship the final version of this book was prepared, as well as the arrangements for the book's publication. I gratefully acknowledge this support and the hospitality of the STS faculty, students and staff.

Finally, I am indebted to Peter Hopkins, past editor-director of Routledge & Kegan Paul, for his willingness to publish this book, and, above all else, to my parents, my dearest friend Michelle Arens, and my brothers, Lewis and Charles, for things that go without saying.

Introduction

A Non-Technical Introduction to Ethnomethodological Investigations of the Foundations of Mathematics through the Use of a Theorem of Euclidean Geometry*

I want to advance two recommendations for the study of mathematicians' work. The first one is this: by returning the mathematical object to its origins within mathematical practice, it is possible to investigate the natural analyzability and natural accountability of mathematicians' work in and as the inspectable, real world practices of professional mathematicians. The second recommendation is that those investigations then offer increasing, and technical, access to the natural analyzability and natural accountability of mathematicians' work as a problem, for mathematical theorem provers, in the local production of social order.

The aim of my talk is to indicate the grounds for these recommendations in ongoing, ethnomethodological studies of mathematicians' work and, in doing so, to offer those studies as a distinctive and vital alternative to classical studies of that work.[1]

To this end, I am going to begin my talk in a somewhat unusual manner: I am going to begin by proving a theorem of Euclidean

---

* This chapter is the text of a talk delivered at the annual meeting of the American Sociological Association in San Francisco in September 1982. It is intended to serve as a non-technical introduction to the remainder of the book. In writing this chapter, I benefited from the extensive and generous help of several people: Charles Livingston spent hours reviewing and helping me develop the mathematics of the paper; Emanuel Schegloff read numerous drafts of the paper, and his suggestions and criticisms were invaluable in articulating and organizing the paper's argument. I am indebted as well to Herbert Enderton for many stimulating conversations and to Michelle Arens, Paul Colomy, Rod Harrison and Anita Pomerantz for their critical suggestions. Responsibility for the material in this chapter is, of course, my own.

geometry. This proof, in that it will recall for us the appearance of mathematical argumentation, will be useful for sketching the current state of studies about mathematicians' work and, then, later, as a means of indicating what some features of the lived-work of mathematics actually look like.

The theorem that I want to prove is this: First, given a circle with center C and an angle α inscribed in it, the measure of that angle's intercepted arc {here }[2] is

defined to be the measure of the angle β that that arc is said to subtend.

The theorem that I am going to prove is that the measure of angle α is 1/2 the measure of angle β, or more properly, that the measure of an inscribed angle is 1/2 the measure of its intercepted arc.

The proof of this theorem goes like this: first, let us distinguish three possible cases: the case where the center of the circle lies on one of the edges of the angle:

the case where the center of the circle is interior to the angle:

and the case where the center of the circle is exterior to the angle:

Let us prove the theorem for {this case} first.

By drawing in the subtended angle and labeling the angles α and β, what we wish to show is that the measure of α is 1/2 the measure of β.

But note that {this line} and {this line} are radii of the circle and that, therefore, they have equal length.

This means that {this triangle} is an isosceles triangle and that, therefore, {this angle} must equal {this angle} so {this angle} is α, also.

Now note that since {this} is a triangle, α plus α plus {this angle} must equal 180°, and since β plus {this angle} make a straight line, β plus {this angle} must also equal 180°. But then the measure of β must equal the measure of α plus the measure of α or that

$$2\,m(\alpha) = m(\beta)$$

or that

$$m(\alpha) = \frac{1}{2} m(\beta)$$

We have finished with case one.

INTRODUCTION

If we turn to {this case} next, we see, by drawing in the diameter of the circle through the vertex of the angle

and the subtended angle

that if we can show that the measure of {α} is 1/2 the measure of {β} and if the measure of {γ} is 1/2 the measure of {δ}

then the measure of the sum of α and γ will be 1/2 the sum of β and δ, or that the measure of the inscribed angle will be 1/2 the measure of the intersected arc. But by our first case, we see that the measure of α is 1/2 the measure of β, and similarly that the measure of γ is 1/2 the measure of δ, so we have finished with case two.

Case three is similar to the last case but a little more difficult to see. Let us first draw in the diameter and the subtended angle.

If we label {this angle} α and {this angle} β, and correspondingly, {this angle} α' and {this angle} β',

INTRODUCTION

then we see that the incribed angle is $\alpha - \beta$ and that the subtended angle is $\alpha' - \beta'$. Then

{'the measure of the inscribed angle $\alpha - \beta$ equals'}

$$m(\alpha - \beta) = m(\alpha) - m(\beta)$$
$$= \frac{1}{2} m(\alpha') - \frac{1}{2} m(\beta') \text{ +'again, by case one'}$$
$$= \frac{1}{2} m(\alpha' - \beta')$$

{'equals the measure of the subtended angle $\alpha' - \beta'$.'}

Thus, for this case as well, the measure of the inscribed angle is 1/2 the measure of its intercepted arc, and we are done.

Now, before undertaking an analysis of some of the work of this proof, I want to use it to sketch briefly the central concerns of previous studies of mathematicians' work.

Throughout its entire history, the study of mathematicians' work has been directly to the solution of three problems: the problem of explicating the notion of a rigorous proof; the problem of characterizing the mathematical object; and the problem of understanding the nature of mathematical discovery.

Let me briefly elaborate on these problems. For provers, a mathematical proof (like the one I just gave) is not a form of rhetorical argumentation; its cogency — whether or not, or in what sense, it actually is the case — is not exhibited as being dependent on the particulars of its authorship or its audience; on the contrary, a proof exhibits, as its witnessible achievement, the demonstrable adequacy and practical objectivity of its reasoning. For provers, a theorem is not true because what it asserts has merely been persuasively argued — the measure of an inscribed angle is not 1/2 the measure of its intercepted arc because I am a good debater — it is true because its proof exhibits it to be so. The problem of explicating the notion of a rigorous proof is that of determining what kind of a thing a proof is that this should be so.

Next, consider the objects that concern a geometer. They are not, for example, the circles and angles that he actually draws on the board, nor are his theorems, in any obvious sense, empirical or empirically-verifiable propositions. Mathematical objects — whether circles and

INTRODUCTION

angles or infinite-dimensional Hilbert spaces and unitary transformations — appear as being disengaged from, and are, as a prover's accomplishment, exhibited as being disengaged from, the actual material circumstances of a prover's examination of them. They are circles qua circles, angles qua angles, and their properties are the properties of circles qua circles and angles qua angles.

The problem of characterizing the mathematical object is the problem of determining what kind of objects a mathematician refers to when he proves theorems about them; what kind of objects are circles qua circles and angles qua angles.

And, finally, let us consider how a theory of the mathematical object affects our appreciation of the nature of mathematical discovery.

I will give two examples. First, if the mathematical object is considered to be an ideal, Platonic object — if the circles and angles of geometry are understood as representing ideal circles and ideal angles — then a theorem of mathematics, like our theorem about inscribed angles, is a description of such objects and a mathematical discovery, in some sense, is an intuition of the transcendental properties of those ideal objects.

If, however, mathematical objects, like circles and angles, are understood as being entirely defined by the structural relationships holding between the symbolic expressions of a purely formal, logistic system, then the achievedly-disengaged objects — like circles and angles — are historical artifacts used for their suggestiveness and for their embeddedness in the practical ways we have of reasoning about them; and a mathematical discovery is a discovery concerning the structure of logical inference which makes its appearance through, but is only incidentally related to, the practical objects of current mathematical interest.

Thus, the problem of understanding the nature of mathematical discovery is seen to be bound up with the problem of characterizing the mathematical object, as, in fact, all three of these problems are tied to one another.

We have, thus, briefly reviewed — in an obviously sketchy manner — the classical problems that have animated studies of mathematicians' work. Now the first thing that I want to call to your attention is a fact massively recognized by practising mathematicians: none of the studies of mathematical rigor, of the nature of the mathematical objects, or of mathematical discovery — whether done by mathematicians reflecting on their own practice, or by philosophers, historians or sociologists — none of these studies are, for the practising mathematician, either instructive in or consequential for his own work practices at the mathematical work-site.

At the work-site, the mathematician does not encounter the problem of mathematical rigor as 'The Problem of Mathematical Rigor,' but as

INTRODUCTION

the endlessly diverse, practical problem of finding and organizing, in their exhibiting and problem-specific material detail, coherent, accountable, mathematical arguments.

At the work-site, the mathematician is not attentive to theories of the mathematical object — Platonic, logistic, or otherwise — his attention is directed to finding and elucidating the properties, for example, of circles qua circles and angles qua angles — their ontology, for practical purposes, being assured in their availability for further precise, rigorous description.

And at the work-site, the mathematician is not interested in a theory of mathematical discovery; he is interested in making mathematical discoveries.

Of course, to those interested in the study of mathematicians' work, these complaints are all too familiar. What can be gained by parading them out?

Let us note the following, curious situation: on every occasion when practising mathematicians come to address seriously, for and among colleagues, the accountability of a piece of mathematical argumentation, and on every occasion when an accountable mathematical argument is discovered, both the specific character of the problem-at-hand and the specific character of that problem's solution arise from within the situated, lived-work of doing professional mathematics — that is, as a work-site phenomenon. But, in contrast, on every occasion when the nature of mathematical activity is posed as its own topic for mathematical, philosophical, historical, or sociological reflection and analysis, the mathematical object is first disengaged from the local work to which its natural accountability and analyzability are integrally tied.

We come, then, to my first recommendation: in order to re-address the classical problems in the study of mathematical practice in a manner that is located within that practice, the mathematical object must first be returned to its origins within that practice itself, and by so returning the mathematical object to those origins, the natural analyzability and accountability of mathematicians' work is capable of being investigated in and as the inspectable, real-world practices of professional mathematicians.

As a way of indicating the basis of this recommendation in current ethnomethodological studies, I want now to return to our proof that the measure of an inscribed angle is 1/2 the measure of its intercepted arc and — given the limitations of time — point out some of the more obvious features of the lived-work of that proof's natural analyzability.

First, let us review the argument that I gave for the case where the center of the circle lay on one of the edges of the angle.

INTRODUCTION

Here, I argued that since {this line} and {this line} are radii, they have equal length, {this triangle} is an isosceles triangle and, therefore, that both {these angles} are α. But, then, since {this angle} plus {this angle} plus {this angle} measure 180°, and since {this angle} and {this angle} measure 180°, the measure of β must be twice that of the measure of α, or that the measure of α must be 1/2 the measure of β.

I Now, the first point that I want to make is this: nowhere in the course of giving this proof did I define the notions of a circle, a triangle, an angle, the measure of an angle, an inscribed angle, or an intercepted arc; and, in fact, nowhere in the proof did it seem that we needed a 'theory' of those objects.[3]

The point to be made is this: in order for us to prove our theorem, we did not need to articulate such definitions. All the properties of angles and measures that we needed were exhibited over the course of giving the proof itself, and, moreover, as fellow provers, what you held me to was the fact that all such properties would be so exhibited.[4] As just one example, consider the fact that in drawing the inscribed angle

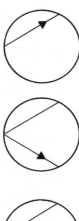

and *then* labeling it α,

8

INTRODUCTION

we are showing that that angle, whatever it is, is related to ⧹this area⧸ between the two drawn chords of the circle. In this way, then, the angles of the drawing become seeably definite; and that this is so then provides for the adequacy of partitioning the inscribed angles into classes determined by the relationship of the angles to the center of the circle: that is, ⧹pointing to the cases drawn on the board⧸ the case where the center lies on one of the edges of the angle, is in the interior of the angle, or is exterior to the angle.

Similar remarks could be made for the definition of the measure of an angle, or an intercepted arc, of triangles, and, in fact, of all the objects of pointed relevance for our proof.

In the course of giving this illustration, another point is illustrated as well — namely, that the figure is temporally drawn and that it takes on its properties by being tied to the embodied action of its depiction.

One sees this property of the proof no more in evidence than when trying to write up the proof on paper. The 'this's' of the drawing — ordered and timed with the embodied pointings, paced writings and accompanying talk — must be removed by locally developed notational devices adequate to the particular problems-at-hand.

Consider for example, labeling the points of the figure

necessary for the proof of our first case. We need just these:

On seeing the figure so labeled, however, the prover sees that ⧹this point⧸ is not labeled. In order to disengage the labeling from its occasioned circumstantiality, a prover will label 'all' the points; moreover, if the prover is going to identify the angle α in the fashion of $\angle$ PQC, then he will want to be able to label β in the same way, thereby removing the need for denoting the angles α and β. Moreover, when the prover comes to label the figure

9

INTRODUCTION

he does so as follows:

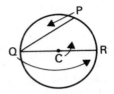

{let C be the center of the circle and let the inscribed angle meet the circle at points P, Q, and R'} thereby 'tracing out' and 'showing again' the inscribed angle, with the center of the circle labeled in such a way so as to disengage it from the angle and being seen as being so disengaged.

Here, of course, a prover will not see this ordering of the work as being essential to the adequacy of the representation: a prover envisions the temporal drawing in terms of the projected gestalt that it is working toward, and he will see beyond another prover's mistakes by rearranging the ordering of the drawing, for himself, so that it is a proper order.

It is the availability of such work that provides the inessential character of a drawing's temporal construction.

But, on the other hand, if a prover does enough things incorrectly — if his writings are not paced with his talk, if he organizes his material presentation improperly — then the naturally analyzable mathematical object will not be exhibited. And that this is so points to the fact that provers work in such a way that this will not be the case. Moreover, from within the production of the naturally analyzable object, the mathematician will use that natural analyzability to argue that the local work of its production had nothing to do with that object's analyzable properties.

Let me draw this set of materials together: if we consider the primordial setting of mathematics to be those occasions when mathematicians, in the presence of one another, work in such a way so as to exhibit to each other the recognizable adequacy of their work, then one of the things that we have seen is that mathematicians work in such a way so as to disengage the mathematical object from the situated work that makes it available and, therein, to disengage that object from the situated work that gives it its naturally accountable properties. Thus, the first major point is that the naturally accountable mathematical object is the local achievement of mathematical provers.

II Let me offer a second collection of observations, these pointing to the work of achieving the objective, transcendental character of the reasoning of our proof.

1 First, let us look at the following drawing for the proof of our first case:

INTRODUCTION

One thing to note about this picture is that there is nothing about the drawing itself that guarantees that these are all the necessary lines needed for the proof. That these are all the necessary lines depends, in fact, on the surrounding course of argumentation from within which the necessary details of the figure take on their relevance specifically for the particular argument being given.

2 A second observation is this: in beginning to prove our theorem, I offered three cases

as being the necessary ones to consider for that proof. Now, without even examining the local work of providing oneself with a scheme of reasoning such that these are established as a disjoint yet exhaustive set of cases, we, as provers, can see that — a priori — there is no particular significance in the fact that inscribed angles can be partitioned in this way — there are other ways of partitioning these angles and, associated with those partitionings, other ways of proving our theorem.[5] The point is that when a prover, as myself, offers these three cases as partitioning inscribed angles into equivalence classes, he is pointing to, and other provers are waiting to see, the course of proving that then follows that exhibits that partition as *a* partition adequate to the proof of the theorem. Moreover, in that a prover has written the cases as first {this one}, then {this one}, and then {this one},

other provers come to look for that temporal course of drawing to be realized in the ensuing proof itself. And in our proof, not only is this ordering preserved, but it reflects a dependency between the cases themselves.

11

INTRODUCTION

3   As one final observation concerning the exhibited reasoning of the proof, we should note that we have proved our theorem for each of the three classes of inscribed angles by proving it for an arbitrary representative of each class. In the first case, {here}

the proof for this picture stood as the proof for all possible inscribed angles in which the center of the circle lay on one of its edges — that is, for all angles like {this} and {this} as well:

Now the thing that gives the arbitrariness of the angles and, thus, that gives the generality of the proof for this class of angles, is the fact that the method of proving exhibits it as an arbitrary angle. That is, the same method of proving our theorem for {this angle} —

by drawing the subtended angle

and reasoning as we did — applies to any other inscribed angle in which the center of the circle lies on one of its edges. Thus, the arbitrariness of the angle — the character of the particular angle as a representative of a class of such angles — is tied to the course of reasoning that exhibits that angle as such an arbitrary one.

INTRODUCTION

In summary, then, these brief observations point to the local, lived-work of producing the practically accountable, practically objective reasoning of our proof.

III The last major point that I wish to make has to do with the larger structural organization of our proof.

Consider the relationship between how a prover comes to find such a proof and the final, materially-presented proof of it. As an example, a prover may, summarily speaking, come to find the relevant cases for proving our theorem about inscribed angles in the following manner: by first reviewing a picture such as {this}

a prover may realize, after trying other schemes, that by drawing the subtended angle

and the diameter through the vertex of the inscribed angle

that if he could prove the theorem for {this case}

then it would follow for {the 'original' case}

INTRODUCTION

Seeing this, and looking at the two cases and searching for an extractable method of classifying them that gives them as two possible cases — like that concerning the relationship of the inscribed angle to the center of the circle — that method then offers both the existence of a third possibility — namely, where the center of the circles is exterior to the angle

— and, therein, offers those three cases as making up all the possible ones.[6]

Now, when a prover comes to write up the proof, he will not detail this work, but will, as I did in proving the theorem earlier, simply offer the three cases as the necessary cases for the proof.

If we put this observation together with the other aspects of the lived-work of proving this theorem that I have detailed, we come to the following point: a prover, in the course of working out a proof, extracts from the lived-work of that proof, the accountable structure of that work — that is, he extracts the specifically remarkable features of the presented proof, and does so against the background of practices that both provide for that structure and that, simultaneously, that structure makes available. A proof, then, is not the disengaged, material argument, but it is always tied to the lived-work of that theorem's particular proving; a proof is this inextricable pairing of the proof and the associated practices of its proving.

Having gone through this review of our proof, I now want to return to my original argument.

Consider that the primordial origins of mathematics, that the generative and sustaining life of the discipline, lies in the coming together of provers who, in the presence of each other, in exhibiting the demonstrable adequacy of a mathematical line of argumentation, exhibit — as their witnessible and recognizable achievement — the practical objectivity of that argument, therein, their presence to it as mathematical provers, and therein, simultaneously, the adequacy of that argument for anyone whomsoever. Consider that mathematics, as a professional discipline, is sustained and renewed, and that it evolves, is revitalized, and is taught on the occasions, and critically on the occasions, when provers come together and do, for and among each other, the recognizably adequate work of doing recognizably adequate mathematics.

Then, let us make one last observation concerning classical studies of mathematicians' work: by addressing the problems of explicating the notion of a rigorous proof, of characterizing the mathematical object, and of understanding the nature of mathematical discovery, classical studies of these problems have sought to explain, and, thereby, have indirectly pointed to and depended on, the great regularities of mathematical practice — namely, that mathematicians regularly produce naturally accountable mathematical proofs; that mathematicians, as their situated accomplishment, endlessly demonstrate the independence of mathematical objects from the local work through which those objects take on their naturally analyzable properties; that mathematicians make recognizable and teachable mathematical discoveries; and that their proofs and discoveries concern, particularly, the naturally accountable objects given by mathematical practice.

We can see, then, in that the originating intention of those studies is to account for the witnessible regularities of mathematical practice, that those studies are, in fact, properly conceived of as theories of social structure and practical action — as theories of how mathematicians, in their daily work, as their daily, specifically- and identifiably-mathematical work, come to produce, maintain, transform and exhibit the observable and identifying orderliness of their practice.

In conclusion, then, we come to the second of my recommendations concerning the study of mathematicians' work. It is this: by seeing beyond the classical studies of mathematicians' work to the produced and witnessible orderliness of mathematical practice that those studies indirectly attempt to explain — that is, by affiliating ourselves with their originating intentions and not with their received histories — and by returning the mathematical object to its origins within the mathematical work-site, it is then possible to take up the study of mathematicians' work not as a means of elaborating familiar and well-known topics, but, instead, as the investigation of the natural analyzability and natural accountability of mathematical practice; to do so as a problem, for provers, in the irremediably local production of social order, and to do so by investigating, as discoverable matters, the natural analyzability and accountability of mathematical practice, in its identifying detail at the work-site, in and as the real-world practices of professional mathematicians.

A Guide to the Reading of this Book

On every occasion when the professional mathematician comes to address seriously, in the company of colleagues, the adequacy, the rigor, the accountability of some piece of mathematics, and on every occasion when an accountable mathematical argument is discovered, both the specific character of the problem-at-hand and that problem's solution arise in the midst of and are made available from within the situated, lived-work of doing professional mathematics. Yet on every occasion in which the origin of mathematical rigor is posed as its own topic for mathematical, philosophical, historical or sociological reflection and analysis, that rigor is disengaged by the analyst from the local work that, in every particular case, provides for its achievement. In this book I raise the question: what constitutes mathematical rigor from within the settings in which that rigor lives as the circumstances of doing professional mathematics? What I propose, and make available as a real-worldly researchable matter, is that mathematical rigor consists of the local work of producing and exhibiting, for and among mathematicians, a 'followable' — and, therein, a naturally accountable — line of mathematical argumentation. If the problem of the foundations of mathematics is understood as the problem of discovering the origins of and analyzing mathematical rigor, then this book can be summarized by saying that it formulates that problem as a problem in the production of social order and solves it in the sense that that formulation is shown to permit the investigation of mathematical foundations in and as the real-world practices of mathematicians engaged in doing professional mathematics.

The book begins with a short introductory chapter giving the necessary definitions that allow me to advance and make precise the claim that conventional studies of the foundations of mathematics are 'classical' studies. In that these studies have themselves become a practiceable mathematical discipline, their legitimacy and authority rest on the fact that the mathematical reasoning that occurs in them is itself mathematically rigorous. The claim that they are 'classical' studies is the claim that the local, order-productive work of an 'ordinary' mathematical argument's natural accountability is irremediably hidden within them in and as the self-same practices of producing naturally accountable mathematics. Thus, the characterization of conventional studies as classical studies is tied to the solution of the foundation problem itself.

As a first step in developing and explicating this claim, and as a means of introducing the reader to the methods employed in foundational studies, Chapter 1 provides a contrast between a very simple mathematical proof and the representation of that proof in a formal logistic system. This contrast is used both to illustrate the inadequacy of such a formal system as a literal description of mathematical practice and to point out the 'reasoner's work' that surrounds even a simple proof and recovers that proof as a naturally accountable mathematical object.

INTRODUCTION

The inadequacy of logistic sytems for descriptively representing mathematical practice is, of course, recognized by mathematicians and mathematical logicians alike; the claim that foundational studies actually concern – and are even primordial to – ordinary mathematical practices rests not on such systems' literal descriptiveness, but on the construction of those systems so as to model supposedly essential features of mathematical practice. Hilbert is generally credited with the idea that the proofs of the consistency of such systems would supply the foundations for those systems, for the mathematical theories that they represent, and, by implication, for the ordinary mathematical practices used in reasoning about those theories. The attempt to develop such systems and to then prove their consistency is generally referred to as the 'Hilbert program.'

It is against the background of this program that the interpretation of Gödel's celebrated incompleteness results, published in 1931, is most often presented. Preceding Gödel's discoveries, one – if not the central – problem of the Hilbert program was to demonstrate the consistency of a formal system 'representing' number theory. Gödel's theorems can be summarized by saying that they showed the impossibility of proving, from within the formal system itself, the consistency of any 'reasonable' formal system strong enough to 'contain' number theory. Thus, Gödel's theorems point to the seeming paradox that, on one hand, the discovery, demonstration and adjunction of rigorous mathematics make up the daily work circumstances of mathematicians and mathematical logicians alike and, on the other, that the independence of mathematical rigor from the lived-work of proving cannot be demonstrated by mathematical methods.

Chapter 2 begins by reviewing this material. The proofs of Gödel's theorems, however, are not proofs given within a formal system, but are 'informal' proofs of everyday mathematics about formal systems. The question is thereby raised: what is it that makes up the rigor of their proofs as proofs of ordinary mathematics? What is it, independently of their metamathematical interpretation, that constitutes the natural accountability of the proofs of Gödel's theorems as 'informal' proofs of everyday mathematics? Thus, Chapter 2 provides an introduction to Gödel's incompleteness theorems and then offers the contrast between the metamathematical interpretation of them and the proposal to study the natural accountability of them in and as the lived-work of their proofs. Part II of the book – divided into chapters 'Gödel Numbering and Related Topics,' 'The Double-Diagonalization/"Proof",' 'A Technical Lemma,' 'A Schedule of Proofs,' and 'A Structure of Proving' – provides an answer to that question by descriptively analyzing the work of proving one of those theorems. Part II is a case study of the origins of mathematical rigor in the work practices of mathematicians.

In 'Gödel Numbering and Related Topics,' I introduce the basic

INTRODUCTION

mathematical apparatus needed for the proof of Gödel's first incompleteness theorem — a Gödel numbering, the definitions of primitive recursive functions and relations, and the definitions of representability and numeralwise expressibility. Although most of the material in this chapter only lays the groundwork for my later analyses, the discussion does point to the various topics that gain importance over the remainder of Part II.

In 'The Double-Diagonalization/"Proof",' the concluding part of the proof of Gödel's theorem is reviewed and then analyzed through a list of fifteen descriptive 'remarks.' These remarks elucidate the ways in which the diagonalization argument and the 'proof' mutually articulate each other and the ways in which that mutual articulation is relied on to construct the diagonalization/'proof.' This analysis of the simultaneous development of a pair — (a mathematical construction/a proof) — clarifies features of some types of mathematical theorem proving but, as the reader will see, it is not intended as a prototypic account of mathematicians' work.

'A Technical Lemma' discusses a second component of the proof of Gödel's theorem — the proof of the lemma that primitive recursive relations are numeralwise expressible in first-order arithmetic. This lemma has the character of being a 'technical residue' of the proof of Gödel's theorem as a whole. In this chapter, I attempt to explain how its character as a 'technical residue' arises from within the work of proving Gödel's theorem itself.

The main body of the proof of Gödel's theorem consists of a sequentially ordered list of functions and relations and the interwoven proofs that those functions and relations are primitive recursive. This list of propositions and proofs will be referred to in the book as the 'schedule of proofs.' The chapter 'Primitive Recursive Functions and Relations' addresses the practical techniques of showing that a function or relation is primitive recursive. The idea of this chapter is to indicate how those practical techniques provide for their own demonstrable adequacy, thereby radicalizing the notion that those techniques are 'grounded.' The material presents a curious backdrop to the work of constructing a naturally accountable schedule of proofs.

'A Schedule of Proofs' begins by outlining a particular schedule of proofs and ends by describing four of that schedule's features in terms of the lived-work of its production. First, a Gödel numbering is shown to be a technique of proving rather than just a correspondence defined between the language of a formal system and the natural numbers. Next, the ordering of the propositions of the schedule is shown to have a 'directed' character. Third, the propositions of the schedule of proofs are shown to be selected and arranged so as to provide for and exhibit the accountable orderliness of the work of producing that schedule. Fourth, I provide some initial materials on the way in which the work

INTRODUCTION

of producing a consistent notation for a schedule of proofs articulates that schedule as one coherent object.

The detailed analysis of the work of proving Gödel's theorem is brought to a conclusion in 'A Structure of Proving.' Here, the problem is raised of specifying what identifies a proof of Gödel's theorem, over the course of its production and review, as a naturally accountable proof of just that theorem. The analysis of the book up to this point allows some immediate constraints to be placed on prospective solutions: primarily, that such a solution cannot be found by reference to an idealized proof or to an idealized collection of mathematical practices. What I propose is that a proof of Gödel's theorem has a 'structure of proving' — that is, that a schedule of proofs is arranged in such a way so as to exhibit the accountable orderliness of the work of producing that schedule, and in that this is so, that schedule simultaneously provides directions for its own recovery as a naturally accountable mathematical object, both in its material detail and as pointing to the diagonalization/'proof' and, therein, to the necessary bridging step of the technical lemma. More generally, the idea that a proof has a structure of proving is that the material proof of a theorem is available in practice as a template of a course of action, that course of action being the work of recovering the proof for which the material presentation stands as its projected, natural and practically adequate account.

The chapter 'A Structure of Proving' brings to an end Part II of the book. Part III, a short concluding chapter, ties together the book's overall argument. Its point is this: even though Part II was restricted to the analysis of the work of proving one particular theorem, the results of that analysis provide for further, similarly detailed investigations of mathematicians' work. What was discovered about the natural accountability of a proof of Gödel's theorem seems to resonate throughout mathematical praxis. In that this is so, the argument turns back on itself: By discovering the origins of mathematical rigor in and as the work of producing naturally accountable, ordinary mathematics, conventional foundational studies, in so far as they are a mathematical discipline, must essentially hide those origins in and as the self-same practices of doing recognizably adequate, rigorous mathematics.

The book concludes with an appendix that points to the consequentiality of my studies of mathematicians' work for the investigation of the relationship between mathematics and theoretical physics.

This, then, outlines the argument of the book. Let me offer some advice about reading it. First, given the claims of the book, it was impossible for me to turn away from the examination of the material detail of mathematical praxis. I have tried to write the book so that its arguments are intelligible over and above the mathematical analysis. The mathematically uninitiated might do well to treat the symbols

INTRODUCTION

as untranslated hieroglyphics. The hieroglyphics should, however, be inspected and, perhaps, by the end of the book, they will begin to take on a fuller life.

A more serious drawback for the mathematically unsophisticated is that I do not think that the descriptiveness and generalizability of the book's findings and the faithfulness (or lack thereof) of my analyses to actual mathematical praxis can be decided without reference to and knowledge of the details of that praxis. Whether or not, in the end, my research brings clarity to the problem of the foundations of mathematical practice, I hope that my book invites the reader into the serious investigation of it.

One last bit of advice. My terminology and definitions (like those of the opening pages) may strike the reader (particularly one unfamiliar with ethnomethodological investigations) as being strange, difficult, obfuscating, or the like. I have tried, on all occasions, to make my terminology descriptively precise. Nevertheless, that precision, or lack thereof, cannot be decided by ruminating over the definitions themselves. The reader is urged to read the definitions but not to dwell on them, to read through the book, but not to dwell on the mathematics, to use the book to find the phenomenon that I am attempting to describe, and then to return to the book to see if I have described that phenomenon accurately or not.

Part I

Introduction. The Phenomenon:
The Existence of Classical Studies of Mathematicians' Work

The aim of this book is to examine the consequentiality of conventional studies of the foundations of mathematics for investigations of the natural accountability of mathematicians' work. In order to get the point of that examination, some preliminary definitions must be given.

The book's motivating concern is to make the natural accountability of mathematics' ordinary organizational objects capable of being investigated as a real-world phenomenon. In particular, mathematical proofs are going to be discussed as being organizational objects. The reference to proofs will always be to the particular here and now proofs that fill mathematical practice. It is to proofs as 'lived work,' to proofs as they are tied to temporally developing and, then, historicized and reviewed displays of written material. The reference will never be to some conception of proofs in general.

The expression 'the natural accountability of a proof' or 'the natural accountability of a proof to its production cohort' will be used to point to the availability of a proof's practical objectivity, of its anonymity as to authorship, of its transcendental character, of its rigor[1] as a witnessed and witnessible feature of the work of proving the theorem, of the work of a proof's situated accomplishment. The notion of natural accountability refers not only to the fact that proofs are done for and done among mathematicians, but that they are done recognizably as proofs; they are produced as the objects that they recognizably are.

Reference to a proof's natural accountability as an *essentially* local phenomenon is to the assertion that a proof's accountability *consists* of the witnessible work of its production.

Last, the term '*classical* foundational studies' will be used to refer to studies of the natural accountability of mathematical proofs for which the essentially local character of that accountability is irremediably hidden in and as the practices that make up the accountability

INTRODUCTION. THE PHENOMENON

of those studies themselves.

The point of this book is this: conventional studies of the foundations of mathematics are, in fact, classical studies.

1 A Review of the Classical Representation of Mathematicians' Work as Formal Logistic Systems

The first task in coming to some understanding of the proposition that conventional studies of the foundations of mathematics are classical studies of mathematicians' work is to review what a formal logistic system looks like and to examine whether or not, or in what sense, such a system is descriptive of the material details of mathematical practice. The 'theory of groups' will serve this comparison well. Group theory should be particularly accessible to formal representation and analysis because it is defined syntactically as the consequences of a given set of axioms. In contrast, a theory like the theory of functions of a real variable is defined in a partially semantic way, partly by axioms and partly as just all things true about functions of a real variable.

A group may be defined in the following way: Let G be a nonempty set and let $*$ be a mapping of $G \times G$ into G; that is, $*$ associates a unique element of G to each pair of elements in G. $x * y$ will be referred to as the product of x and y in G. G is a group if and only if the following conditions hold:

(G1) $\forall x, y, z \in G \quad x * (y * z) = (x * y) * z$ (associative law)

(Read: 'for all x, y, and z in G, the product of x with the product of y and z is equal to the product of the product of x and y with z.')

Now, let $\exists$ mean 'there exists.' The next axiom posits the existence of an identity element:

(G2) $\exists e \in G$ such that $\forall x \in G \quad e * x = x * e = x$.

In words, G2 says that there is an element e of G such that the product of e with any other element of G is always equal to that other element.

Before stating the third and last axiom in the definition of a group, it is possible to prove the theorem – to be identified as Theorem T – that the identity element of a group posited in axiom G2 is unique. This is an example of a, perhaps, ultimately simple proof, and I will use this proof as a means of developing and illustrating the contrast between actual mathematical practice and the representation of that practice within a logistic system. To prove the theorem, consider first that to say that the identity element of a group is unique is to say that if there were two identity elements, they would be equal. The proof of the theorem is given in one line and goes like this: if there were two identity elements e and e', then

$e = e' * e = e'$.

The last axiom in the definition of a group posits the existence of 'inverses':

(G3) Let e be the identity element posited in G2. Then $\forall x \in G \; \exists y \in G$ such that $x * y = e = y * x$.

In order to develop our comparison between actual mathematical practice and the representation of that practice within a logistic system, the rudiments of mathematical logic will first have to be developed. The real aim of the underlying logistic system of a formal theory is to provide an exact, syntactically specified notion of a 'formal proof.' Reference to a formal proof (a proof within a formal system) will be distinguished from references to the 'informal' proof of ordinary mathematics by speaking of the former, as is sometimes done, as 'deductions.' Theorems of a formal theory, as opposed to theorems of ordinary mathematics (including those about such a formal theory), will be denoted as *theorems*. All the details cannot be given here; the following discussion is only intended to give a flavor of what is involved.

The immediate object of this presentation is to represent the proof of Theorem T in a formalization G* of group theory. The first step in developing G* is to specify its 'language' by providing a collection of signs to serve as the primitive symbols of the language and by giving exact formation rules for building up its syntactic objects – like 'variables,' 'terms,' and 'well-formed formulas (wffs)' – from these primitive symbols through the operation of concatenation. In general, the parameters of a language will change from theory to theory, as in the case of G* where there is only one predicate letter ✱, to be interpreted as * in G, in addition to the two-place predicate for equality.

The underlying logistic system of a formal theory should contain as *theorems* all wffs (well-formed formulas) that are 'tautologically true.' This can be accomplished by the following three axioms:

$A \supset . B \supset A$

$A \supset [B \supset C] . \supset . [A \supset B] \supset [A \supset C]$

$\sim\sim A \supset A$

where A, B, and C are any wffs.

Additional axioms for first-order predicate calculus are

$\forall x A \supset S^x_t A \mid$

where $S^x_t A \mid$ indicates the process of 'substitution,' x is an individual variable, and t is a term 'free for x in A,' and

$\forall x[A \supset B] \supset . A \supset \forall x B$

where x is an individual variable having no 'free' occurrences in A.

A two-place predicate, represented by the standard equality symbol, and the axioms

$\forall x[x = x]$

$x = y \supset . A(x,x) \equiv A(x,y)$

where A is any wff obeying a certain restriction, introduces the notion of equality into the logistic system.

Finally, the (non-logical) axioms for the formal theory of interest (in this case, elementary group theory G*) are added to the logical axioms. The reason for speaking of G* as being 'elementary' group theory will be explained below. With $\underline{*}$ as a three-place predicate symbol defining the group operation in G*, the axioms for G* are the following:[1]

(G*1) $\forall x \forall y [\exists z (\underline{*}(x,y,z)) \land$

$\forall u \forall v [\underline{*}(x,y,u) \land \underline{*}(x,y,v) \supset u = v]]$

(G*2) $\forall x \forall y \forall z \forall s \forall t \forall u [\underline{*}(x,y,r) \land \underline{*}(r,z,s) \land$

$\underline{*}(y,z,t) \land \underline{*}(x,t,u) \supset s = u]$

(G*3) $\exists e [\forall x [\underline{*}(e,x,x) \land \underline{*}(x,e,x)] \land$

$\forall x \exists y [\underline{*}(y,x,e) \land \underline{*}(x,y,e)]]$.

G*1 formalizes the notion that * is a function on $G \times G$; G*2 corresponds to the associative law G1, and G*3 to both the existence of an identity G2 and the existence of inverses G3.[2]

The final component of our logistic system is the rules of inference. For our system, we take these to be *modus ponens,*

from A and $A \supset B$, to infer B,

and generalization,

if x is an individual variable, from A to infer $\forall xA$.

Modus ponens is used in the following fashion: let the fact that A is a *theorem* be written $\vdash A$. Similarly, if $A \supset B$ (read: A 'implies' B) is a *theorem*, write $\vdash A \supset B$. An application of *modus ponens* allows the line $\vdash B$ to be written. In the vernacular, if A implies B and A, then B. An application of generalization would allow the line $\vdash \forall xA$ to be written after the line $\vdash A$.

At long last, the notion of a deduction (in G*) – the counterpart of the notion of a proof (in group theory) – can be defined. A deduction of A_n is a finite sequence $A_1, \ldots, A_n$ of wffs of the language of a particular theory (G*) such that for each i, $1 \leqslant i \leqslant n$, one of the following three conditions holds:

(1) A_i is an axiom,
(2) for some r, s $<$ i, A_r is of the form $A_s \supset A_i$,
    (i.e., A_i results from A_r and A_s by an application *of modus ponens*), or
(3) A_i is of the form $\forall xA_s$ for some A_s, s $<$ i
    (i.e., A_i results from A_s by an application of generalization).

A wff A is a *theorem* (of G*) when A is the last wff of some deduction (in G*). As is customary, a turnstile $\vdash$ ($\vdash_{G*}$) will be used to indicate that the formula following the turnstile is a *theorem* (a *theorem* of G*).

The preceding presentation has glossed over many features of the definition of a formal logistic system and formal theory. The material in hand does allow the reader to get an idea of how the theorem T of group theory that was proved earlier can be deduced as a *theorem* T in G*. The deduction that follows is, in fact, only an abbreviation of a full derivation of T. The steps of a full deduction can be recovered through the use of the justificatory comments that follow each line of the argument. An actual deduction of $T*$ in G* would be written without the use of abbreviations, comments, or metatheorems. In practice, as will become apparent, this is never done except in the most elementary of cases.

The theorem T that a group has unique identity element may be 'translated'[3] into G* as the sentence T,

$$\exists e \forall e'[\forall x[\underline{*}(e',x,x) \land \underline{*}(x,e',x)] \supset e = e'].$$

A 'deduction' of this sentence in G* follows:

Let A be the wff $\forall x(\underline{*}(s,x,x) \land \underline{*}(x,s,x)) \land$

$\forall x \exists y(\underline{*}(y,x,s) \land \underline{*}(x,y,s))$, and let B be the wff

$\forall x(\underline{*}(e',x,x) \land \underline{*}(x,e',x))$. Then

$G*1; G*2; A; B \vdash \underline{*}(e',s,s)$	(universal instantiation from B; detachment)
$G*1; G*2; A; B \vdash \underline{*}(e',s,e')$	(universal instantiation from A; detachment)
$G*1; G*2; A; B \vdash \underline{*}(e',s,s) \wedge \underline{*}(e',s,e') \supset s =$	(universal instantiation twice in $G*1$; detachment; universal instantiation two more times)
$G*1; G*2; A; B \vdash s = e'$	(*modus ponens* from last line and conjunction of first two lines)
$G*1; G*2; A \vdash B \supset s = e'$	(deduction theorem)
$G*1; G*2; A \vdash \forall e'(B \supset s = e')$	(generalization on a deduction from hypotheses)
$G*1; G*2; A \vdash \exists e \forall e'(B \supset e = e')$	(existential generalization)
$G*1; G*2; \exists eA \vdash \exists e \forall e'(B \supset e = e')$	(existential instantiation, since s does not occur in either $G*1$, $G*2$, or $\exists e \forall e'(B \supset e = e')$)

Since $\exists eA$ is the same as $G*3$ and $\exists e \forall e'(B \supset e = e')$ is the same as T, we have shown

$\vdash_{G*} T.$

In mathematics instruction, after the definition of a group is given and some of a group's elementary[4] properties are proved – as, for example, the uniqueness of a group's identity element – the topic that is ordinarily introduced next is that of subgroups. A 'subgroup' is a subset of a group G that is itself a group with respect to the group operation (∗) already present in G. The attempt to represent the study of subgroups in a formal system increases the complexity of that formal system tremendously. Whereas, previously, 'quantification' ($\forall x$; 'for all x') was used only with respect to individual variables – variables which range over the elements of a group in their intended interpretation – the formal study of subgroups necessitates quantification over predicate symbols ($\forall A$) as well as over individual variables. It becomes necessary to express things like 'all subsets of a group with a certain structure ...' and 'if A is a subgroup of G, then ... ,' and a first-order theory is not adequate to this task. Parenthetically, this is the reason for referring to G∗ as elementary (that is, first-order) group theory; it is a theory that is restricted to studying relationships between

MATHEMATICIANS' WORK AS FORMAL LOGISTIC SYSTEMS

the elements of a given group and not between subsets of a group.

The material presented so far offers a direct contrast between the proof of a theorem in ordinary mathematics and the representation of a proof (if feasible) in a formal system. Hopefully that contrast will lend cogency to the following series of claims.

1 The representation of in-progress research within a formal system is treated by mathematicians engaged in that research as being completely irrelevant to their research.

2 The translation of current work into the deductive schema of a formal theory is never contemplated nor seen as being feasible.

3 The representation of work in a formal logistic system is never carried out as a check on either one's own or on one's colleagues' work.

4 Although the ability to represent mathematical reasoning in a formal system (as, for example, axiomatic set theory) is spoken of as the ultimate source of the legitimacy of the reasoning of a mathematical argument, such a check is always proposed as an 'in principle' possibility and is, in practice, completely infeasible.

5 The study of formal logistic systems are not found by mathematicians to be illuminating of the natural technology of their situated inquiries into mathematical structure. Such studies are completely unenlightening as to the origins of mathematical creativity and discovery.

6 The representation of an argument in a formal system is not, in any ordinary sense, descriptive of the work of the original proof. On the contrary, in an immediate sense, it is obfuscating of that work.

This last point can be elaborated through the example of the proof $e = e' * e = e'$. In the first place, the attempt to formalize that proof — to deduce the translation of the theorem T in G* — makes the proof of that theorem problematic. Secondly, it makes problematic the obvious thing being proposed by the single line $e = e' * e = e'$, that it is proposed in a single line, and that such economy is essential to the obviousness of that proposal. The claim that needs to be advanced is this: the line $e = e' * e = e'$ is, in fact, *read* as a proposal; the proposal can be articulated as the claim that '$e' * e$' can be read in two specific ways; when that line is read as a proposal, that reading initiates a 'small-time' inquiry that is almost immediately satisfied in the way that it asks that the axioms for a group be reviewed as to their 'meaning'; that 'meaning' comes to reside in the seeable thing proposed about '$e' * e$'.

7 A ubiquitously recognized feature of mathematics instruction is that the occasioned need to make such 'small-time' inquiries a theoretical enterprise turns the teaching of mathematics into a hopeless enterprise.

2 An Introduction to Gödel's Incompleteness Theorems:

Their Metamathematical Interpretation Contrasted with the Proposal to Study Their Natural Accountability in and as the Lived-Work of Their Proofs

The overall concern of this book is to examine the bearing of conventional foundational studies on the investigation of the natural accountability of mathematicians' work. In the previous chapter, the formal representation of mathematical practice was compared to that practice itself, and it was argued that that representation was not, in any straightforward way, descriptive or illuminating of mathematicians' work. However, the guiding inspiration of foundational studies is not that the formal systems that are studied are literally descriptive of mathematical practice, but that they represent the essential features of that practice, and that by proving things about formal systems, things are actually being proved about the essential, innate, or inherent structure of mathematics itself. According to this account, the natural accountability of mathematicians' work is but an intimation of the transcendental character of mathematical reasoning.

In Part II of this book, the feasibility of this program of conventional foundational studies will be addressed through an examination of a proof of one of Gödel's incompleteness theorems. The argument that will be made is *not* that a proof of Gödel's theorem does not prove what others have claimed it to prove; instead, the origins of the rigor of a proof of Gödel's theorem will itself be examined and the claim advanced that that rigor consists of its local work. Thus, this argument points to the primordial character of the activity of doing mathematics over some conception of mathematics-in-itself.

After completing this argument, we will be in a position to return to our initial proposal that conventional foundational studies are, in fact, classical studies.

The present chapter is intended mainly as an introduction to Gödel's incompleteness theorems, to their metamathematical interpretation, and to the contrast between the interpreted content of Gödel's theorems and the natural accountability of proofs of Gödel's theorems as rigorous proofs of just those theorems.

The theorem of Gödel that I am going to discuss concerns any 'reasonable'[1] first-order axiomatization of number theory. My presentation begins, however, with a particular axiomization of that theory which will be referred to as elementary (Peano) arithmetic and denoted by the letter P. P is obtained from the logistic system of the last section by removing the axioms of group theory, adding axioms for elementary arithmetic, and adjusting the parameters of the language by eliminating the three-place predicate symbol $\stackrel{*}{=}$ and adding a constant symbol O, a one-place function symbol S (for the successor operation), and the two-place function symbols $\pm$ and $\dotdiv$. For the parts of the proof of the theorem that will be of interest here, an actual set of axioms need not be given, and I refer the reader to conventional sources. A schema of mathematical induction must be included in the axioms of P if the theorem is to be 'embedded' in P itself.

As our terminology indicated, the *theorem* of G*, that was deduced earlier was an example of a *theorem* 'proved' *within* a formal system. In contrast to this, the theorem of Gödel to be discussed is a metatheorem; it is a theorem that concerns the structure of P as a formal system. The theorem is not deduced in P, but is proved as a theorem of ordinary mathematics about P.

In order to state the theorem, the notion of the consistency of a formal system must be introduced. P is said to be consistent if there is no wff A of P such that both A and $\sim A$ ('not'–A) are *theorems* of P. The import of this notion is that if, for some wff A, both $\vdash_P A$ and $\vdash_P \sim A$ then by the tautology $\sim A \supset . A \supset B$, all the wffs of P would be *theorems*. If this were so, the claim that mathematics concerns 'platonic objects' (in this case, number theory) would (in the case of number theory) be shattered; number theory would have no definite set of properties. Furthermore, the elemental character of number theory would multiply this result throughout mathematical practice. On the other hand, the proof of the consistency of P could be interpreted as saying that ordinary mathematical reasoning does, in fact, prove things about mathematical objects.

Although this line of reasoning may well have been at the basis of their research, the formalists (referring principally to Hilbert and his co-workers) at least claimed to understand mathematics as itself being investigations within formal systems that, in practice, were treated and studied in an informal manner. To the formalists, P made explicit what was already implicit in ordinary mathematical investigations of number theory. The proof of the consistency of P was itself taken as supplying the absolute grounding of that theory and, therein, the adequacy of ordinary mathematical reasoning about it.

The discussion of the philosophical consequences of Gödel's theorem will be facilitated by first speaking about the theorem that

INTRODUCTION TO GÖDEL'S INCOMPLETENESS THEOREMS

Gödel *almost* proved: if P is consistent, then there is a sentence J of P such that neither J nor $\sim J$ are deducible in P. By extending the previously noted generality of Gödel's theorem to this one, the theorem says that in any 'reasonable' first-order axiomatization of number theory (including, for example, P with J added as an additional axiom), if that system is consistent, there will be some sentence S of that system such that neither S nor $\sim S$ are deducible in it. This is often expressed by saying that 'number theory' is essentially incomplete. Finally, if it were possible to supplement this theorem with the fact that one of the sentences S or $\sim S$ is intuitively true of number theory, the theorem could be interpreted as saying that there is no adequate formalization of number theory.

Let us contrast the theorem[2] that Gödel actually did prove to the one just mentioned. Gödel's theorem says that there is a sentence J of P such that (1) if P is consistent, J is not deducible in P, and (2) if P is ω-consistent, then $\sim J$ is not deducible in P. This is the theorem that will occupy us, and it will be referred to either as 'Gödel's first theorem' or simply as 'Gödel's theorem.'

ω-consistency (the definition of which will not be needed for now) is a stronger condition than consistency. Correspondingly, by weakening Gödel's theorem, one obtains the theorem that says that if P is ω-consistent, there is a sentence J such that neither J nor $\sim J$ is a *theorem* of P. Since, in practice, number theory is assumed to be both consistent and ω-consistent, the philosophical impact of Gödel's theorem is left unmarred. The thing that was seen to be so remarkable about the theorem was that it was proved by first indicating how such a sentence J could be constructed and then by showing that the assumption of either J or $\sim J$ led to a contradiction. Because Gödel's 'formally undecidable' sentence J is not deducible in P, it can also be seen to be intuitively true.[3]

The tremendous significance of this result for traditional studies of the foundations of mathematics was eclipsed (at least philosophically) by the significance of a consequence of just the first part (1) of Gödel's theorem, a consequence indicated by Gödel himself in the same paper in which his first result appeared.

To state this result and briefly indicate its significance, the technique of 'Gödel numbering' will have to be introduced. As will be discussed later, it is possible to assign 'Gödel numbers' to each symbol, formula and deduction in P. Letting g denote this assignment, $m = g(M)$ would be the Gödel number of the formula M. Now, let $k(m) = S^m(0)$ be the numeral in P corresponding to the number m. Then $k(g(M)) = k(m) = S^m(0)$, and $k(g(M))$ would be the numeral in P corresponding to the Gödel number of the formula M.

Using a Gödel numbering, it is possible to 'arithmetize' Gödel's first theorem and prove *this* *theorem* as a deduction in P. To do this,

33

a formula $Cons_P$ is constructed in P which can be interpreted as saying that P is consistent, and, similarly, a formula $\sim Th(k(j))$ is constructed, where $j = g(J)$, and which can be interpreted as saying 'k(j) is not a *theorem* of P' — k(j) being the numeral representing the Gödel number of the sentence J which can be interpreted as saying 'I am not a *theorem* of P.'

Gödel argued that the 'arithmetization' of the first part of his theorem — $Cons_p \supset \sim Th(k(j))$ — is deducible in P. The implication of this result is this: If $Cons_p$ is a *theorem* of P, then so is $\sim Th(k(j))$ by *modus ponens*. Next, $\sim Th(k(j))$ can be shown to imply J. But if J is a theorem of P, by the actual proof of Gödel's first theorem (and as suggested by the interpretation of J itself), P can be shown to be inconsistent. Thus, for P to be consistent, $Cons_P$ cannot be deducible in P. This is Gödel's second incompleteness theorem.

Generalizing the statement of this theorem, as with Gödel's first theorem, to any 'reasonable' axiomatization of number theory, Gödel's second theorem showed that if number theory was consistent, the arithmetical statement of its consistency could not be deduced as a *theorem* of number theory, or, in the vernacular, that one could not prove the consistency of number theory in number theory itself. This result can be generalized even further. In any formal system, like axiomatic set theory, strong enough to define the natural numbers and have the axioms of P as *theorems*, a similar statement concerning that system's consistency can be deduced. Since set theory is understood to be strong enough to define all of mathematics, the consistency of mathematics, via Gödel's theorem, was interpreted as being unprovable within mathematics itself. Thus, the transcendental character of mathematical reasoning — the disengagability of that reasoning from the actual, temporally-situated, circumstantial, at-the-board or with-pencil-and-paper work of doing mathematics — was found to be incapable of mathematical demonstration.

I have developed the statement and interpretation of Gödel's theorems to the extent that I have in order to give the reader a sense of the way in which those theorems (and the theorems of conventional foundational studies in general) can be seen to be descriptive of mathematical practice. An oddity of this interpretive descriptiveness is also illustrated by Gödel's work. On one hand, his work was tremendously consequential for the development and practice of foundational studies; on the other, it had no effect whatsoever on the way in which ordinary mathematical investigations were actually conducted. Contrarily, the mathematicalization of foundational studies was greatly accelerated.

This curious situation points to a paradox at the heart of the attempt to interpret the results of conventional foundational studies in terms of mathematical practice. The paradox is this: the daily and identifying work of professional mathematics consists, for mathematicians and

INTRODUCTION TO GÖDEL'S INCOMPLETENESS THEOREMS

mathematical logicians alike, of the work of discovering and establishing demonstrably rigorous mathematical arguments.

Part II of this book initiates the investigation of how, in fact, this is so, through a descriptive review of a proof of Gödel's first theorem itself. The argument to be made is that the rigorous character of the reasoning of that proof resides in the local work of the proof itself.

As the reader will recall, P is being used to designate a particular axiomatization of elementary Peano arithmetic, and the theorem that says 'there is a sentence J such that if P is consistent, J is not deducible in P, and if P is ω-consistent, then $\sim J$ is not deducible in P' is being referred to as Gödel's first theorem.

I want to now outline the argument that I will make. In the proof of his theorem, Gödel introduced two major techniques of meta-mathematical research: 'Gödel numbering' and 'the double-diagonalization procedure.' Together, these techniques give the work of proving his theorem a directional character which is actualized as a programmatically accomplished, intrinsically sequentialized series of tasks. This way of proving will be called a 'structure of proving.' I begin my analysis by introducing the notions of Gödel numbering and double-diagonalization and by indicating that they do provide for the proof of Gödel's theorem as a programmatic series of tasks. A review of some of the work of that program will then elucidate three major claims: (1) the proof-specific structure of proving Gödel's theorem is made available in and as the local work of actually proving the theorem; the structure of the work and the work itself are mutually identifying as a course of action; (2) the naturally accountable objectivity, rigor or truthfulness of that work — whether the proof's demonstrable clarity, its flawed and inadequate character, or its evident erroneousness — consists of that identification; and (3) that (1) and (2) are so points to the organizational or 'social' character of mathematicians' work.

Part II A Descriptive Analysis of the Work of Proving Gödel's First Incompleteness Theorem

3 Gödel Numbering and Related Topics:

Background Materials for a Proof of Gödel's Theorem

In this chapter, I give the mathematical definitions that are needed to discuss Gödel's theorem and indicate, in a preliminary fashion, how these definitions are coordinated over the course of proving Gödel's theorem. Primitive recursive functions and relations, numeralwise expressibility, and representability will be given exact definitions,[1] and a specific Gödel numbering will be partially described and illustrated. The meaning of these definitions in and as the detailed, in-situ work of proving Gödel's theorem will be developed in the chapters that follow.

Although primitive recursion functions and relations are of great importance to the proof of Gödel's theorem, only the definitions of them will be needed for now.

A numerical function is, for some $m > 0$, a function $f : N^m \to N$, where N stands for the natural numbers (i.e., $0, 1, 2, \ldots$; the numbers of number theory). A numerical relation R is, for some $m > 0$, a subset of N^m. The numerical functions Z (the zero function), S (the successor function), and I_i^m (the projection functions)

$Z(x) = 0$

$S(x) = x + 1$

$I_i^m(x_1, \ldots, x_i, \ldots, x_m) = x_i$

are referred to collectively as the *initial functions*. If h is an n-place numerical function and $g_1, \ldots, g_n$ are m-place numerical functions, then the m-place function f defined by

$f(x_1, \ldots, x_m) = h(g_1(x_1, \ldots, x_m), \ldots, g_n(x_1, \ldots, x_m))$

is said to be obtained from $g_1, \ldots, g_n$ and h by *substitution*. If g is an m-place numerical function and h is an $(m + 2)$-place numerical function, then the $(m + 1)$-place function defined[2] by

GÖDEL NUMBERING AND RELATED TOPICS

$$f(x_1, \ldots, x_m, 0) = g(x_1, \ldots, x_m)$$
$$f(x_1, \ldots, x_m, S(y)) = h(x_1, \ldots, x_m, y, f(x_1, \ldots, x_m, y))$$

is said to be obtained from g and h by *primitive recursion*.

The *primitive recursive functions* are inductively defined as only those functions obtained from the initial functions by a finite number of substitutions and primitive recursions. From the fact that this is an inductive definition, it follows that if the initial functions are shown to possess some property,[3] and if that property is shown to be preserved by substitution and primitive recursion,[4] then every primitive recursion function possesses that property.

Let the characteristic function of an m-place numerical relation be denoted by K_R,

$$K_R(x_1, \ldots, x_m) = \begin{cases} 1 \text{ if } (x_1, \ldots, x_m) \in R \\ 0 \text{ if } (x_1, \ldots, x_m) \in N^m - R. \end{cases}$$

An m-place numerical relation is a *primitive recursive relation* if its characteristic function is a primitive recursive function.

The next definitions that need to be introduced are those of 'numeralwise expressibility' and 'representability.' The idea of numeralwise expressibility is that if a relation W is numeralwise expressible, then there is a formula W of P that 'defines' that relation in P. Technically, an m-place numerical relation W is *numeralwise expressible* in P if and only if there is a wff $W(x_1, \ldots, x_m)$ of P with m free variables such that the following conditions hold:

(1) if $(a_1, \ldots, a_m) \in W$, then $\vdash_P W(k(a_1), \ldots, k(a_m))$

(2) if $(a_1, \ldots, a_m) \notin W$, then $\vdash_P \sim W(k(a_1), \ldots, k(a_m))$

where, as before, k(a) is the numeral of P corresponding to the number a.

For a numerical function, the notion corresponding to the numeralwise expressibility is that of representability. An m-place numerical function f is *representable* in P if and only if there is a wff $F(x_1, \ldots, x_m, x_{m+1})$ of P with m+1 free variables which has the properties that, for any natural numbers $a_1, \ldots, a_{m+1}$,

(1) if $f(a_1, \ldots, a_m) = a_{m+1}$, then $\vdash_P F(k(a_1), \ldots, k(a_m), k(a_{m+1}))$,
and

(2) $\vdash_P \exists! x_{m+1} F(k(a_1), \ldots, k(a_m), x_{m+1})$.

('$\exists! x \ldots$' is read 'there exists a unique x such that ...')

The idea of both numeralwise expressibility and representability is that some formula of P 'performs a "role"' in P similar to the 'role' performed by the original relation or function in number theory.

For the discussion of Gödel's theorem the relationship between representability and numeralwise expressibility that is needed is this:

if R is an m-place numerical relation and if its characteristic function K_R is representable in P by $K_R(x_1, \ldots, x_{m+1})$, then R is numeralwise expressible in P by $K_R(x_1, \ldots, x_m, S(0))$.[5]

Later, I will discuss the fact that all primitive recursive functions are representable in P. From this fact, from the fact that primitive recursive relations have, by definition, primitive recursive characteristic functions, and from the results just mentioned it follows that primitive recursive relations are numeralwise expressible in P.

In the next section, the numeralwise expressibility of primitive recursive relations will be given greater emphasis than the representability of primitive recursive functions. Gödel numbering permits syntactic properties of P to be associated with numerical relations. These relations can be shown to be primitive recursive; hence, there are formulas of P corresponding to those relations that numeralwise express them. *In this manner*, certain formulas of P can be seen to formulate properties of P's own syntax.

Later, in the discussion of 'the technical lemma,' the importance of representability will come to the fore.

The last major definition to be introduced is that of Gödel numbering itself. The predominant way of introducing this topic is simply to give a specific assignment of numbers to the primitive symbols, expressions,[6] and finite sequences of expressions of P (that assignment being defined as a Gödel numbering) with the anticipation/foreknowledge that the work of proving Gödel's theorem will come to show that that particular assignment fulfills the requisite properties of its intended use. Typically, some of the 'computability' properties of that numbering are explicitly mentioned. I begin my presentation in a similar manner by illustrating a conventional type of Gödel numbering. Afterward, I will indicate the practical requirements on a Gödel numbering for the proof of Gödel's theorem.

A Gödel numbering is specified by first assigning numbers to the primitive symbols, then, using these, to the expressions, and, then, using these, to the sequences of expressions of P. Consider, for example, the string of symbols (x''') which will be abbreviated x_3 and used syntactically as an individual variable. Under the following assignment of numbers

(...... 3
) 5
' 7
x 9

(x''') can be written as an ordered 6-tuple of numbers (3, 9, 7, 7, 7, 5). Now consider the prime numbers arranged as an increasing series:

GÖDEL NUMBERING AND RELATED TOPICS

2, 3, 5, 7, 11, 13, 17, 19, 23, 29, 31, . . . ,
or symbolically, p_1, p_2, p_3, ... (p_0 will be defined as 1.) To the 6-tuple (3, 9, 7, 7, 7, 5) associate the number $p_1^3 \cdot p_2^9 \cdot p_3^7 \cdot p_4^7 \cdot p_5^7 \cdot p_6^5$ or $2^3 \cdot 3^9 \cdot 5^7 \cdot 7^7 \cdot 11^7 \cdot 13^5$. This is the Gödel number $g(x_3)$ of x_3.

Next, for finite sequences of expressions, Gödel numbers can be assigned as follows: Consider $\sigma = x_4, x_3$ as a sequence of expressions (and not the expression obtained by concatenating the symbols of x_4 with those of x_3). Then the Gödel number of σ is $2^{g(x_4)} \cdot 3^{g(x_3)}$. As there is some ambiguity is using g to denote the Gödel numbering function, I will use g only for the Gödel numbers of expressions, as with $g(x_3)$.

Finally, with this numbering, it is theoretically possible to work backwards and determine if a given number is a Gödel number and, if so, what symbol, expression or sequence of expressions corresponds to it. Given any number, one can, in principle, uniquely factor it as a product of primes and, from this decomposition, reconstruct the number's corresponding 'element' of P if it has one. Thus, a prime factorization of $g(x_3)$ would yield $g(x_3) = 2^3 \cdot 3^9 \cdot 5^7 \cdot 7^7 \cdot 11^7 \cdot 13^5$, from which (3, 9, 7, 7, 7, 5) and then (x''') could be recovered.

The major definitions that have been introduced so far — primitive recursive functions and relations, numeralwise expressibility, representability, and Gödel numbering — can be brought together in a list of three related practical requirements on a Gödel numbering. By 'practical requirements,' I mean that a Gödel numbering needs to fulfill these requirements in order that a proof of Gödel's theorem be carried out with sufficient detail to exhibit that proof's rigorous character.[7]

Let Lg(P) denote the set of all primitive symbols, expressions, and finite sequences of expressions of P. A Gödel numbering is an assignment of numbers to the elements of Lg(P) that (minimally) obeys the following restrictions:

(1) That assignment is one-to-one — each element of Lg(P) is assigned only one number, and each number has at most one corresponding element in Lg(P). Technically such an assignment is an 'injection' from Lg(P) into N; informally, it is a 'renaming' of the elements of Lg(P) as natural numbers. Under the association of elements of Lg(P) with natural numbers established by a particular Gödel numbering, to speak about a specific number can be intuitively interpreted as speaking about the element of Lg(P) that corresponds to it, and to speak about an element of Lg(P) can be interpreted, vice versa, as speaking about the number corresponding to it.

In the presence of such an assignment, let $G(x,y)$[8] denote the numerical relation 'x is the number of a deduction of the wff which results when k(y) is properly substituted for a specified individual variable z in the wff with number y.'

GÖDEL NUMBERING AND RELATED TOPICS

(2) The numerical relation G(x,y) defined under the correspondence established by a particular assignment of numbers to the elements of Lg(P) can be shown to be primitive recursive.

(3) That assignment allows the 'proof' that G(x,y) is primitive recursive to be accomplished by 'demonstrating' that G(x,y) is the result of the construction of a chain of functions and relations, built by substitution and primitive recursion, from the initial functions and from previously established primitive recursive functions and relations of that chain. These demonstrations are understood to represent the explicit display of a sequence of primitive recursive functions, built stepwise according to the definition of primitive recursive functions, for either the functions themselves or for the characteristic functions of the relations involved. Later, this notion will be made precise as a 'formal proof' and the explicit construction as a 'formal construction sequence.'

The work-related point of this restriction is that the practically available rigor of the main body of work of proving Gödel's theorem 'consists' of that sequentialized series of 'demonstrations.'[9]

In order to give the philosophical significance of this restriction, the following result must be stated:[10] the representability of primitive recursive functions in P is proven by giving explicit wffs representing the initial functions and by showing how explicit wffs representing a function obtained by substitution or primitive recursion can be constructed from the explicit wffs representing the functions used in that substitution or primitive recursion. The philosophical point of restriction (3) is that, following from this method of proof, there is a projectable construction in P[11] of a wff $G(x_1, x_2)$ corresponding to the projectable formal construction in number theory and, thus, in principle the formally undecidable sentence of Gödel's theorem is capable of being constructed in P and materially exhibited as a formally undecidable sentence.

Despite their theoretical multiplicity, there is essentially only one commonly used Gödel numbering and only several explicitly described types of such numberings in the literature. The reason for this is that a Gödel numbering must demonstrably fulfill the practical requirements listed over the course of a proof of Gödel's theorem. Theoretically, that the construction of restriction (3) can be made is the heart of the proof of Gödel's theorem; in principle, most of the work of that construction is trivial; practically, the construction is a matter of extended labor and care. The presence of an explicitly described, proof-specific Gödel numbering in a proof of Gödel's theorem is essential for the practical objectivity of the temporally situated and developing, detailed work that makes up the demonstrations of restriction (3).

The point of listing these three restrictions is this: the explicit statement of these practical requirements — as opposed to concealing them in an abstract definition of Gödel numbering — begins to locate the local work and the situated, discovered-in-course, material detail of proving Gödel's theorem.

A severe misconception will result if the detailed co-ordination of definitions pointed to here is understood as a transcendental property of the (platonic) objects or properties defined by them — it turns Gödel's discoveries into an heroic achievement, not a human one. Rather than the various definitions fitting together as pieces of a jigsaw puzzle, the definitions are, in a natural way, tied to the existence of a Gödel numbering. As we shall see, Gödel numbering is not just a numbering; it is a-numbering-as-a-technique-of-proving. In conjunction with the diagonalization procedure/'proof' to be discussed next, Gödel-numbering-as-a-technique-of-proving provides an endogenous organization for the work of proving Gödel's theorem. The proof-specific Gödel numbering (as a technique of proving) exhibits — over the course of the work of proving Gödel's theorem, as the accomplishment of that work of proving — a technical compatibility with the practical techniques of working with primitive recursive functions and relations. That 'compatibility' — as a name for the thing that the work of proving Gödel's theorem makes practically available — makes up the practical objectivity of that work itself.

4 The Double-Diagonalization/ 'Proof':

Features of the Closing Argument of a Proof of Gödel's Theorem as Lived-Work

The sequence of definitions through which the formally undecidable sentence of Gödel's theorem is constructed will be referred to here[1] as the 'double-diagonalization procedure' or as one or more of the possible variants of this name.

In the previous chapter it was noted that the numerical relations corresponding under a specific Gödel numbering to certain syntactic properties of P are primitive recursive and, therefore, are numeralwise expressible in P. In this way,[2] formulas of P can be interpreted as 'speaking about' P itself. Heuristically, the double-diagonalization procedure constructs a wff which 'says' 'I am not a *theorem* of P.' The fundamental point of this construction is that the formula so obtained has the syntactic property that neither it nor its negation are deducible in P if P is consistent or ω-consistent, respectively.

By the '"proof"' of Gödel's theorem, I refer to the statement of Gödel's theorem and the argument following that statement (placed after the diagonalization procedure) that the sentence constructed through the diagonalization procedure has the syntactic properties just mentioned. The 'proof' speaks of one aspect of the practical organization of the proof of Gödel's theorem through which the accountable structure of the work of that proof is exhibited. The proof of Gödel's theorem and the 'proof' are two distinct, but overlapping sets of practices.

This chapter initiates the task of elucidating the endogenously organized work of the naturally accountable, practically objective proof of Gödel's theorem. First, I recall for the reader the material presentation of the diagonalization procedure and the 'proof'. Some summarizing comments are made about the work of constructing the undecidable sentence in terms of the jointly articulating character of the diagonalization procedure and the 'proof'. The pairing of the diagonalization procedure/'proof' points to the character of that pair as a 'closing argument.'

THE DOUBLE-DIAGONALIZATION/'PROOF'

For the 'proof' that follows, we will need the definition of ω-consistency, and I begin by stating it.[3] P is said to be ω-consistent if for no wff $F(x)$, both $\vdash_P \exists x F(x)$ and $\vdash_P \sim F(k(0)), \ldots, \vdash_P \sim F(k(n)), \ldots$ for all $n \in N$. Intuitively, ω-consistency means that when any wff $F(x)$ is interpreted in number theory, it cannot be simultaneously false for all natural numbers and true for some natural number. If P is ω-consistent, then it is consistent or, equivalently, if P is not consistent, it is also not ω-consistent. This is so because if P is not consistent, all formulas of P are deducible in P.

For the double-diagonalization procedure, we may proceed as follows: Let $\text{sub}_{g(x_2)}(x, y)^4$ be the Gödel number of the wff which results from 'properly' substituting the numeral $k(x)$ for the individual variable x_2 in the wff with Gödel number y. I.e., $\text{sub}_{g(x_2)}(x, y) = g(S^{x_2}_{k(x)} A|)$ where $g(A) = y$. Let $\phi(u) = \text{sub}_{g(x_2)}(u, u)$. Next, let $\text{ded}(x, y)$ denote the relation 'x is the Gödel number of a deduction in P of the wff with Gödel number y.' $\phi(u)$ and $\text{ded}(x, y)$ can be shown to be a primitive recursive function and relation, respectively, from which it follows[5] that the relation defined by $G(x, u) \Leftrightarrow \text{ded}(x, \phi(u))$ is also a primitive recursive relation. Let $G(x_1, x_2)$ be a wff numeralwise expressing G in P.

Define I as $\sim \exists x_1 G(x_1, x_2)$ and let $g(I) = i$.

Define J as $\sim \exists x_1 G(x_1, k(i))$ and let $g(J) = j$.

Finally, note that $\phi(i) = j$, that is, that the Gödel number of the wff resulting from the substitution of $k(i)$ for x_2 in the wff with Gödel number i (i.e., I) is the Gödel number of J.

Gödel's First Incompleteness Theorem. (1) If P is consistent, J is not deducible in P. (2) If P is ω-consistent, $\sim J$ is not deducible in P.

Proof of (1): If J is deducible in P, that is, if $\vdash_P J$ or

(*) $\qquad \vdash_P \sim \exists x_1 G(x_1, k(i))$,

then, for some r, $\text{ded}(r, j)$. Since

$\text{ded}(r, j) \Leftrightarrow \text{ded}(r, \phi(i)) \Leftrightarrow G(r, i)$,

the numeralwise expressibility of G gives

$\vdash_P G(k(r), k(i))$.

Thus,

(**) $\qquad \vdash_P \exists x_1 G(x_1, k(i))$.

But (*) and (**) are contradictory, so if P is consistent, J cannot be deducible in P.

Proof of (2): Suppose $\sim J$ is deducible in P. Then $\vdash_P \sim J$ and

(***) $\qquad \vdash_P \exists x_1 G(x_1, k(i))$,

THE DOUBLE-DIAGONALIZATION/'PROOF'

But if $\vdash_P \sim J$, then ded(r, j) cannot hold for any $r \in N$, since, otherwise, we would have $\vdash_P J$ (by definition of ded(r, j) above), contradicting regular consistency. But if ded(r, j) does not hold for any $r \in N$, G(r, i) $\Leftrightarrow$ ded(r, ϕ(i)) $\Leftrightarrow$ ded(r, j) also does not hold for any r. Using the numeralwise expressibility of G, one obtains

$$\vdash_P \sim G(k(1), k(i))$$

.
.
.

(****) $$\vdash_P \sim G(k(n), k(i))$$

.
.
.

for all $n \in N$. (***) and (****) contradict the assumption of ω-consistency. Thus, if P is ω-consistent, $\sim J$ is not deducible in P.

The following list of features of the diagonalization and 'proof' gives some indication of the objectifying work of their production. However, the intention of this chapter is not to provide a detailed analysis of that work, but to point to the fact that the diagonalization and 'proof' form an intrinsically related pair of objects and that together, in terms of the larger structure of proving Gödel's theorem, they constitute an 'ending' for the proof of that theorem.

(1) The diagonalization procedure and 'proof' do not form a pair of 'a discovery'/'the proof' but are the practically accountable, proof-specific reconstruction of something previously 'known' to be capable of such a reconstruction.[6]

(2) A produced feature of the diagonalization procedure is its detailed notational consistency. Consider the selection of the variable for which the substitution of k(i) is made, indicated in the function $sub_{g(\ )}$. In the diagonalization above, x_2 was used, making that selection consistent with $G(x_1, x_2)$, the wff numeralwise expressing G(x, u) in P. Thus, $G(x_1, x_2)$ could be quantified over x_1, and x_2 was left as the variable free for the substitution of k(i). Had x_1 been chosen as the variable for substitution, the formula $G(x_n, x_1)$ would have appeared, calling forth an explanation of the ordering of the variables in G and, perhaps, an explanation of the availability of x_1 as a free variable in a wff numeralwise expressing G.

There are other ways of bringing about this notational consistency, and I mention two of them. G and ded could be redefined by reversing the order of the variables in their predicative notation, thus obtaining G(u, x) $\Leftrightarrow$ ded(ϕ(u), x). $G(x_1, x_2)$ could then be used to numeralwise express G, x_2 would be the variable for quantification, and x_1 would be used as the variable for the substitution. The benefit of this procedure is that the intention of the choice of the first individual variable goes unnoticed, whereas the choice of the second individual variable raises the question of why the second was selected and not the first. Another technique of bringing about a consistent notation is to camouflage the problem through the use of meta-variables like x, y, and z.

(3) Associated with the problem of developing a consistent notation is the problem of developing *a* notational system adequate to the circumstantial, proof-specific details of the diagonalization and 'proof'. Different authors develop different notations for the diagonalization, and the two problems — that of developing a notational system and that of developing a consistent notational system — arise together, are a thematic concern of working out the material presentation of the diagonalization and 'proof', and are solved simultaneously. The notion of a 'consistent' notation is, in practice, equivalent to a notation that, in its orderly presentation, provides sufficient analytic detail for the diagonalization and 'proof'.

(4) The existence of a proof-specific notation for the diagonalization procedure is essential to the diagonalization's adequacy as the sequence of definitions leading to the observation $\phi(i) = j$ and the 'proof'.

(5) That the mathematician is able to disengage the practically objective, material presentation of J from the real-time, real-world work of producing an orderly, proof-specific, analytically adequate notation is the accomplishment of that work. In the presence of an adverse arrangement of notation — discoverable in the course of developing the diagonalization and 'proof' — the mathematician, as a matter of familiar practice, immediately begins to reconstruct that notation so as to exhibit its orderly, proof-specific adequacy. Only when the work of such a reconstruction is repeatedly frustrated does the claimed adequacy of the potential demonstration begin to be called into question.

(6) The reconstruction of the diagonalization procedure and 'proof' mutually articulate each other, not just as finished objects, but as temporally developing, mutually elaborating constructions. A person engaged in proving Gödel's theorem discovers, over the course of writing out the diagonalization and the first part of the 'proof', the necessary technical details that permit their joint development. The projected first part of the 'proof' is consulted to elicit the necessary structure of the diagonalization; the second half of the 'proof' is recovered with the undecidable sentence in hand from the very way that ω-consistency is formulated to permit that 'proof'.

Typically, the work of this mutual, temporally developing articulation is summarized by the mathematician by speaking of the material display of the diagonalization as the 'remembered thing' that it is then, retrospectively, seen to be.

(7) The diagonalization procedure and 'proof' are recovered together,

THE DOUBLE-DIAGONALIZATION/'PROOF'

but as a pair of distinct objects. The work of the production of that pair maintains, as an integral feature of that work, the distinction between the diagonalization and the 'proof'. The two are constructed as the pair 'the diagonalization'/'the proof' as the practically accountable structure of the work of their construction.

(8) The diagonalization and 'proof' are organized in such a way as to evince 'a proper way of doing.'

Consider, for example, the placement of the crucial observation that $\phi(i) = j$. $\phi(i) = j$ is not proved, or, rather, is proved like the proof $e = e' * e = e'$ discussed in Chapter 1. $\phi(i) = j$ is a proposal that points to the thing that ϕ can be understood as saying. The factual character of $\phi(i) = j$ resides in and as its placement in the sequence of definitions preceding it, that placement making transparent the temporally extending reasoning behind the recovery of the meaning of $\phi(i)$. If the equation $\phi(i) = j$ were placed in the 'proof' just before the point where it was first needed, that assertion would exhibit, instead of its factual character, its status as requiring the work of its justification. 'One' would have to, and would, 'go back' and review the sequence of definitions of the diagonalization to recover the reasoning that gives $\phi(i) = j$ its factual character. In this way, the diagonalization/'proof' would be (mentally) reorganized into a proper ordering of topics.

(9) The 'proper' order of the items making up the diagonalization and 'proof' — and of a mathematical proof in general — are made available through the work of their material presentation. The need for certain results, notational corrections, abbreviations, restrictions, and the like is discovered as a circumstantial feature of the thing that needs to be done immediately.

At the blackboard, at the time of the apparent disarray, the reorganization of the work is immediately indicated, commonly by boxing off the needed result with chalk lines from the rest of the work on the board and by offering elaborating comments or by reviewing the board so as to indicate via the notational apparatus available there, how the appropriate emendations could be made. The produced objectivity of both the need and the adequacy of the emendations and corrections allows those emendations and corrections to be interpreted as having come before the work that indicated their need.

The continual reorganization of the material displays of mathematicians' work is an omnipresent feature of mathematical practice. The immediate point is that (i) the organization of the work of proving is an abiding concern in that the practical objectivity of that work recognizably depends on its orderliness and (ii) the practically accountable, analytically adequate orderliness of the work is a produced feature of that work itself.

(10) The discoverable orderliness of the 'proof' as a natural course of reasoning is tied to the way in which the diagonalization procedure

produces the sentence J. The diagonalization provides a pattern that recurs in the 'proof', and, therein, immediately presages the 'proof' as an ordinary course of reasoning. J is constructed from $G(x_1, x_2)$, which is itself obtained from ded and ϕ through the use of the numeralwise expressibility of primitive recursive relations. In the 'proof', the same type of construction sequence is used to prove the contradictory character of asserting either J or $\sim J$. The assertion of either J or $\sim J$ is first translated into an assertion of assertions concerning the Gödel numbers i and j, ded, and ϕ. A theorem or theorems are then obtained by numeralwise expressing those assertions in terms of $G(x_1, x_2)$.

(11) The orderliness of the diagonalization procedure is not inherent in that procedure, but is tied to the things that must be done with it in the 'proof'.

(12) The discovered orderliness of the 'proof' makes up its practical objectivity as a 'proof.' Each proposition of the 'proof' appears from within the work of recovering its reasoning as the natural, sequentially next proposition, 'seeably'[7] factual in and as its placement in the sequence, and, therein, making available the purported claim that the chain of propositions proves the theorem.

(13) What is being recovered in the mutual articulation of the diagonalization and 'proof' are the material details through which the orderly way of working through the diagonalization and 'proof' is made apparent. The work of recovering the diagonalization/'proof' is the work of formating[8] the diagonalization and proof so as to provide for that orderly course of reasoning.

(14) The existence of a formally undecidable sentence of P is fundamentally tied to the existence of a sentence of P adequate to the methods of the 'proof'.[9] What the diagonalization procedure does is to embed the proposition that there exists *a* sentence with certain undecidability properties into the techniques of proving that have been developed over the course of the work leading to the diagonalization and 'proof'.

(15) The structure of the diagonalization/'proof' as a formated pair is oriented to as part of the natural organization of the work of proving Gödel's theorem. Once enough apparatus is developed to permit the construction of J, the 'hard-labour' of the proof is over. From within the work of proving Gödel's theorem, the diagonalization procedure/ 'proof', as a formated pair, are oriented to and worked for as a 'closing argument.'

5 A Technical Lemma:

A Lemma Used in the Proof of Gödel's Theorem; Its Origins as a Technical Residue of the Work of Proving Gödel's Theorem within that Self-Same Work

The numeralwise expressibility of the relation G defined above is of crucial importance for the diagonalization/'proof' of Gödel's theorem, both for the construction of the sentence J and for the proof of that sentence's peculiar syntactic properties. However, the proof that G is numeralwise expressible in P emerges not as part of the diagonalization/ 'proof', nor as an aspect of G particularly, but as a part of the larger structure of proving Gödel's theorem. As we shall see later, the proof of Gödel's theorem develops through the sequentialized demonstrations of the primitive recursiveness of various numerical functions and relations corresponding to syntactic features of P under a specific Gödel numbering. These demonstrations lead to the construction of G as a primitive recursive relation. The numeralwise expressibility of G is obtained by proving that *all* primitive recursive relations (and, hence, G) are numeralwise expressible in P.

Although the fact that primitive recursive relations are numeralwise expressible in P is, thus, of critical importance, the proof of this fact has the character of being a 'technical residue' of the work of proving Gödel's theorem. The aim of this chapter is to indicate why this is so. That it is so will allow us, in the chapters that follow, to give greater attention to the places where the practically available rigor of the proof of Gödel's theorem actually lies.

To develop the argument of this chapter some background information must be given. In his original paper,[1] Gödel did not use the formal system P,[2] but worked within a logistic system similar to that of Whitehead and Russell's *Principia Mathematica* (PM).[3] There are two immediate differences between PM and P: (1) PM has only one nonlogical 'constant,' indicated simply through predicative notation and interpreted as saying, in the case of $F(x)$, that x has the property F, whereas P has four constants 0, S, $+$, and $\cdot$; and (2) PM has an infinite number of predicate variables of all 'types,' whereas P has only individual

51

A TECHNICAL LEMMA

variables, that is, variables of type 1. Furthermore, in PM, it is possible to quantify over a variable of any type, permitting the expression $\forall F(F(x) \equiv F(y))$ to be a wff of PM. The strength of a logistic system like PM is illustrated by the fact that equality, introduced in P through explicit axioms, can be defined in PM; in PM, $x = y$ is a notational abbreviation for the wff $\forall F(F(x) \equiv F(y))$. In that PM has axioms of comprehension,[4] all the properties of equality can be deduced from this definition as *theorems* of PM.

For the discussion that follows, the Peano axioms must also be introduced.[5] Briefly, the Peano axioms assert the existence of a set N and a function S, S: N → N, such that S is injective but not surjective (an axiom of infinity) and such that if any subset L of N has the property that $L \not\subset S(N)$ and $S(L) \subset L$, then $L = N$ (the induction axiom). The collection of all true propositions following from these axioms is referred to as Peano arithmetic. It can be shown that there is a unique isomorphism between any two mathematical theories that satisfy the Peano axioms, and together with a certain amount of informal set theory, these axioms are sufficient for the development of the theory of natural numbers.

A formal statement of the induction axiom given above minimally requires a second-order logic (that is, a logistic system allowing quantification over predicate variables) and cannot be given in a first-order system like P. Basically, P is an approximation of Peano arithmetic in a first-order logistic system. In PM, on the other hand, the Peano axioms can be formulated and a set $\bar{N}$ and a relation $\bar{S}$ can be defined such that within $\bar{N}^6$ the Peano axioms hold. The pair $(\bar{N}, \bar{S})$ so defined is referred to as an 'inner model' of Peano arithmetic in the logistic system of *Principia Mathematica*. That PM was constructed so as to have such an inner model, and the elaborateness of PM, was, in good measure, determined by the attempt to produce such a model without incorporating any potential contradictions into its construction.

Gödel himself did not introduce Peano arithmetic as an inner model in PM but, instead, superimposed a set of axioms for Peano arithmetic on the basic logistic system of PM. This procedure insures the existence of an infinity of 'individuals' within a specified type, an issue of elaborate concern in the original formulation of PM. Gödel also did not use the ramified type theory of PM, but a theory of simple types alone. The simplification thus introduced into PM made Gödel's formal system — for the purposes of his incompleteness theorems — essentially an axiomatic formulation of second-order Peano arithmetic that included an axiom of comprehension. In the following, I will refer to Gödel's system — with some terminological inaccuracy — as (formal) Peano arithmetic or PA.

For the proof that primitive recursive relations are numeralwise expressible in PA, it is first necessary to show how to define wffs in PA

A TECHNICAL LEMMA

that 'correspond' to the initial functions and that 'correspond' to functions obtained by substitution or primitive recursion from numerical functions with previously established 'corresponding' wffs. The notion of 'correspondence' is elaborated — and the proof of the numeralwise expressibility essentially completed — by showing (1) that the 'associated' wffs so defined act as functions in PA[7] and (2) that if $f(a_1, \ldots, a_m) = b$, then $\vdash_{PA} F(k(a_1), \ldots, k(a_m), k(b))$, where F is the wff of PA 'corresponding' to f.[8] These conditions are, of course, those of the (strong)[9] representability of a function in PA.

Let us suppose that 'corresponding' wffs of PA could be found for the primitive recursive functions in the sense that the construction of these wffs in PA mimic the construction of their associated functions in number theory. Then (2) and the restriction of (1) to numerals (that is, representability) would intuitively follow in that for any $(m+1)$-tuple of numerals $(k(a_1), \ldots, k(a_m), k(b))$, the substitution of these numerals in F would yield a variable-free formula[10] of PA and, hence, a formula either deducible in PA or one whose negation is deducible in PA. If $\vdash_{PA} F(k(a_1), \ldots, k(a_m), k(b))$, then for any other $(m+1)$-tuple $(k(a_1), \ldots, k(a_m), k(c))$, $c \neq b$, $\vdash_{PA} \sim F(k(a_1), \ldots, k(a_m), k(c))$.

Continuing in this intuitive fashion, consider the construction of the wffs to be associated with primitive recursive functions. The wff $\bar{S}$ corresponding to S, either by construction in PM or by explicit postulation in PA, acts as a function in PM or PA, respectively. The wffs representing the projection functions and the constant function Z are fairly obvious ($x_i = y$ and $\forall y \sim S(y, x)$, respectively), and that they act as functions to be expected, $\forall y \sim S(y, x)$ by definition and $x_i = y$ by the properties of equality. Also easily defined is the wff corresponding to a function obtained by substitution from numerical functions previously shown to have corresponding wffs. The only potential difficulty arises in the case of a function obtained by primitive recursion. Yet, in informal set theory, the Peano axioms assure that a function obtained by primitive recursion is well-defined. In that PA and PM contain a formalization of the Peano axioms, the construction of a wff 'corresponding' to a function obtained by primitive recursion can mimic the definition of such a function in number theory, and the proof that such a function is well-defined can similarly be translated into PA or PM, implying that that wff acts as a function in PA or PM.

The preceding remarks are only heuristic, and what remains is the work of materially proving that such 'corresponding' wffs can be defined and that (1) and (2) hold for them. That material presentation does, however, follow the heuristic sketch just given. Once the corresponding wffs are found, they are seen to be 'natural' definitions; the proofs of (1) and (2) follow by induction on the length[11] of the construction of a primitive recursive function in number theory; in both

A TECHNICAL LEMMA

cases, the proofs for the wffs corresponding to the initial functions are elementary, and the proofs for the wff corresponding to a function obtained by substitution require only slightly more ingenuity. In the case of (1), the proof for the wff associated with a function obtained by primitive recursion is technically difficult, but mimics the proof of informal set theory that a function so obtained is well-defined, and, finally, the proof of (2) primarily consists of checking to see that the right formula holds for the right numerals. This synchrony between a natural way of proving and the orderliness of the eventual proof as the material realization of that way of proving makes up the straightforward character of the proof of the numeralwise expressibility of primitive recursive relations in PA. Gödel himself noted that the actual proof was direct, if somewhat laborious, and gave only a brief indication of it in his paper.[12]

One last aspect of the proof of the numeralwise expressibility of primitive recursive relations in PA needs to be introduced — that of its 'role' as a 'structure of accountable inference' in the larger proof of Gödel's theorem. As we shall see later, a proof-specific Gödel numbering is essential to the proof of Gödel's theorem in that the work of demonstrating the primitive recursiveness of the needed numerical functions and relations involves that numbering's explicit presence. In contrast to this use of Gödel numbering, the lemma of the numeralwise expressibility of primitive recursive relations and its proof are detached from the detailed work of the material presentation of the larger proof. Instead — as with the passage from the statement $G(r, i)$ to the *theorem* $\vdash_{PA} G(k(r), k(i))$ — the lemma provides only an accountable structure for the writing of various lines of the diagonalization and 'proof.'[13]

The material presented so far can be summarized in three points. First, the proof of Gödel's theorem develops by demonstrating that the numerical functions and relations corresponding under a specific Gödel numbering to syntactic features of PA have the structure of primitive recursive functions and relations. That G is a primitive recursive relation is the aim of those demonstrations, and that G is numeralwise expressible in PA is a result of the fact that it is primitive recursive. Thus, the question of the numeralwise expressibility of G emerges as part of the larger structure of proof of Gödel's theorem. Second, the proof of the numeralwise expressibility of primitive recursive relations is straightforward in the sense that the intuitive, 'natural' way of proving that proposition is realized as the orderliness of the proof. In that this is so, the proof of that proposition becomes a technical exercise. And third, the role of the proposition and its proof in the proof of Gödel's theorem is only to provide an accountable structure for writing various lines of that proof and is not involved with the detailed, material presentation of the proof of Gödel's theorem itself.

A TECHNICAL LEMMA

Once the proof of Gödel's theorem was given for PA (interpreted as PM), the importance of that theorem was immediately established. The proof of the theorem for the formal system P is essentially the same as that for PA, except for the question of the numeralwise expressibility of primitive recursive relations in P. Although the initial functions and functions obtained by substitution should be (and are) representable in P, P does not have the formal apparatus that PA has to mimic definitions by primitive recursion. Thus, the problem of proving the numeralwise expressibility of primitive recursive relations in P devolves on finding some way of translating the definition of a function by primitive recursion in number theory into a formal construction of a wff representing that function in P from the wffs representing the functions used in that definition.

This problem can be made more explicit. Suppose that the $(m + 1)$-place function f is obtained by primitive recursion from the m-place numerical function g and the $(m + 2)$-place numerical function h, and furthermore, suppose that $f(a_1, \ldots, a_m, b) = c$ for some numbers $a_1, \ldots, a_m$, b and c. Then, from the definition of primitive recursion, there exists a sequence of numbers $n_0, \ldots, n_b = c$ such that

$n_0 = g(a_1, \ldots, a_m, 0)$

$n_1 = h(a_1, \ldots, a_m, 0, n_0)$

$n_2 = h(a_1, \ldots, a_m, 1, n_1)$

.
.
.

$n_{b-1} = h(a_1, \ldots, a_m, b-1, n_{b-2})$

$c = n_b = h(a_1, \ldots, a_m, b, n_{b-1})$.

If a function obtained by primitive recursion from functions representable in P is itself to be representable in P, then it is necessary to show (1) that there is a numerical function β such that, for every finite sequence of natural numbers $n_0, \ldots, n_i, \ldots, n_b$, there exists natural numbers r and s for which $\beta(r, s, i) = n_i$ for $0 \leq i \leq b$, and (2) that there is a wff representing this function in P. Once this wff is obtained, it can be used to define the values that the wffs representing g and h must assume in order to construct a wff representing a function obtained by primitive recursion. The work that remains is the technically involved proof that the wff so obtained actually represents the original function in P.[14]

That such a function β exists is a known result of (ordinary) number theory. However, suppose for a moment, that the definition of β could not be given in such a way so as to show that it is representable in P. In the first place, that function certainly would be representable in

A TECHNICAL LEMMA

the system P with an additional function symbol and axioms for exponentiation.[15] Secondly, if β were not representable in P — and, by implication, if functions obtained from representable functions by primitive recursion were not representable in P — then P might not, retrospectively, be considered an adequate formalization of number theory.

Let us now return to the theme of this chapter: that the proof of the numeralwise expressibility of primitive recursive relations — and, more particularly, that the proof of the numeralwise expressibility of the relation G — is a 'technical residue' of the proof of Gödel's theorem. To speak of that proof as a 'technical residue' is to make reference to the recognized coherence and structure of the work of proving Gödel's theorem from within that work. In the chapters that follow, this endogenously produced orderliness of the proof of Gödel's theorem will be identified as a 'structure of proving.' Anticipating the description of this 'structure of proving,' the material in this chapter indicates why, as part of that 'structure of proving,' the proof of the numeralwise expressibility of primitive recursive relations has the character of being a 'technical residue.' In the way that the problem arises and is formulated within the proof of Gödel's theorem, in the way that it consists of working out the technical details of its argument, in that its proof is anticipated in PM, in that its proof is an indication of the adequacy of P as a formalization of number theory, and in that its 'role' in the proof of Gödel's theorem is not intrinsic to the material presentation of the proof, but only supplies an accountable structure for certain parts of that presentation all point to the intuitive accuracy of the formulation of that proof as a 'technical residue.'

6 Primitive Recursive Functions and Relations:

An Initial Discussion of the Irremediable Connection between a Prover's Use of the Abbreviatory Practices/Practical Techniques of Working with Primitive Recursive Functions and Relations and the Natural Accountability of a Proof of Gödel's Theorem

So far, two organizational components of the work of proving Gödel's theorem have been identified: an ending argument consisting of the construction of the diagonalization/'proof' and the proof of a 'technical lemma' asserting the numeralwise expressibility of primitive recursive relations in P. The major remaining task of the following chapters is that of disclosing the orderliness and rigor of the proof of Gödel's theorem as the local work of their joint accomplishment. In order to do this, one more aspect of that proof needs to be examined — namely the abbreviatory practices used in working with primitive recursive functions and relations. The present chapter reviews these practices and gives a brief indication of their role in the proof of Gödel's theorem.

The reader will recall that primitive recursive functions are defined inductively as only those functions obtained by a finite number of substitutions

$f(x_1, \ldots, x_m) = h(g_1(x_1, \ldots, x_m), \ldots, g_n(x_1, \ldots, x_m))$

and primitive recursions

$f(x_1, \ldots, x_m, 0) = g(x_1, \ldots, x_m)$

$f(x_1, \ldots, x_m, S(y)) = h(x_1, \ldots, x_m, y, f(x_1, \ldots, x_m, y))$

from the initial functions: the zero function, the successor function, and the projection functions —

$Z(x) = 0$,

$S(x) = x + 1$, and

$I_i^m(x_1, \ldots, x_i, \ldots, x_m) = x_i$,

respectively. The reader will also recall that a primitive recursive relation is a numerical relation R whose characteristic function K_R is a primitive

57

recursive function. K_R serves to 'identify' the members of R; K_R can assume only the values 0 and 1, and if $K_R(x_1, \ldots, x_m) = 0$, then $(x_1, \ldots, x_m) \notin R$, and if $K_R(x_1, \ldots, x_m) = 1$, then $(x_1, \ldots, x_m) \in R$.

For the discussion that follows, several more definitions will be needed. First, a *formal construction sequence* for a primitive recursive function f is defined as a finite sequence of functions $f_0, \ldots, f_n$ with the properties that (1) for every $0 \leq i \leq n$, f_i is either an initial function or is obtained from the functions of the collection $\{f_0, \ldots, f_{i-1}\}$ by substitution or primitive recursion and (2) $f = f_n$. The definition of primitive recursive functions guarantees that every primitive recursive function has at least one (and hence, many) associated formal construction sequences. A *formal proof* that a given numerical function f is a primitive recursive function consists of the explicit display of a formal construction sequence and the demonstration that f satisfies the defining equation(s) of the last function of that sequence. Finally, the expression 'seen/shown' will be used to refer to the fact that what is seeably true about a mathematical object *is* seeably true in that that 'seeing' is embedded in the practical techniques of making the thing seeably true, accountably so.

As has already been pointed out, the body of the proof of Gödel's theorem consists of a series of propositions and proofs that certain functions and relations are primitive recursive. The proofs of that series are never formal proofs in the sense just defined; instead, they are understood by mathematicians as representing such formal proofs through the use of various abbreviatory practices. In what follows, I will briefly examine some of the features of these informal proofs and the abbreviatory practices associated with them.

In the first place, explicit construction sequences for functions are never actually displayed. The informal proof that a function or relation is primitive recursive is placed within a sequentialized series of similarly informal demonstrations, and the intervening steps between the previous construction and the demonstration in question are assumed to be easily recovered from the formula(s) defining the new function or relation. For example, once addition and an operation like 'limited subtraction,' symbolized by $\dotdiv$,

$$x \dotdiv y = \begin{cases} x - y \text{ if } y \leq x \\ 0 \quad \text{ if } x < y \end{cases}$$

have been shown to be primitive recursive, the material proof that the equality relation on N^2 is primitive recursive is simply the equation $K_=(x, y) = 1 \dotdiv ((x \dotdiv y) + (y \dotdiv x))$. The adequacy of this proof consists of the work of finding the seeable/showable structure of $K_=$ as the result of a series of substitutions and of computing the value of $K_=$ when $x = y$ and when $x < y$.[1]

PRIMITIVE RECURSIVE FUNCTIONS AND RELATIONS

The example of $K_=$ illustrates a second abbreviatory practice used in working with primitive recursive functions and relations: the check or argument that the equation(s) demonstrating the primitive recursiveness of the function or relation in question does define that function or relation is not given, but is partially left to the reader and partially understood to be fulfilled by the construction of the equation(s) itself. Thus, the need for the calculation indicated above — that the equation $K_=(x, y) = 1$ when $x = y$ and $= 0$ when $x \neq y$ — is left to the reader, and in the case of the proof that addition is a primitive recursive function, the equations

$$x + 0 = x$$
$$x + S(y) = S(x + y)$$

are either assumed to be recognized as the recursive definition of addition or assumed to be understood in the sense that since ordinary addition satisfies these equations, and since the equations define a unique function, that function must be ordinary addition. As in the example of $K_=$, the calculations that are needed are generally obvious, but they are of interest here in that once 'one' writes the formula for $K_=$ (or for term(x) below), 'one' immediately checks the formula against the criteria used in obtaining it, and the exclusion of that work from the material presentation of the informal proof is part of the work of trivializing the omnipresence of such local, situationally-specific work as constituents of the practically adequate formula itself.

A third abbreviatory practice concerns the conventions for naming functions and relations. In the case of $K_=$, rather than specifically naming the numerical function $(x, y) \mapsto (x \dotminus y) + (y \dotminus x)$ and the constant function SZ, the function $K_=$ is identified through its value at $(x, y) - K_=(x, y) = 1 \dotminus ((x \dotminus y) + (y \dotminus x))$. The numerical function of addition is identified simply as $x + y$; multiplication, as xy; 'limited subtraction,' as $x \dotminus y$; and so forth. The initial functions Z, S, and I_i^m are written as 0, $x + 1$, and x_i, respectively.

Fourth, the material demonstration of the primitive recursiveness of a function is never an explicit display of a definition by substitution or primitive recursion. Instead, unspecified, informal ways of writing epitomizing demonstrations are employed. In the elementary case of addition, the recursive equations

(*)
$$x + 0 = x$$
$$x + S(y) = S(x + y)$$

are 'understood' to stand proxy for the primitive recursive equations that can be constructed from them as

$$\eta(x, 0) = I_1^1(x)$$

59

PRIMITIVE RECURSIVE FUNCTIONS AND RELATIONS

$$\eta(x, S(y)) = (S(I_3^3)) (x, y, \eta(x, y))$$

which, in turn, seeably/showably compute the same values as (*) and, hence, define addition of natural numbers — $\eta = +$. In the case of primitive recursion relations, equations for their characteristic functions — except in a few elementary cases like equality — are not given. Instead, various ways of constructing primitive recursive relations from previously established primitive recursive functions and relations are used to establish the primitive recursiveness of the relation in question.

Finally, the example just given points to a fifth abbreviatory practice: the use of theorems showing that if certain functions and relations are primitive recursive, then a particular way of defining a function or relation from them will also be primitive recursive. Constructing a formula in such a way that it seeably/showably fits the template provided by such a theorem constitutes an adequate demonstration that that function or relation is itself primitive recursive. An elementary example of such a theorem is the proposition that if S and T are m-place primitive recursive relations, then 'S and T' will be one also. (Proof:[2] $K_{S \text{ and } T}(x_1, \ldots, x_m) = K_S(x_1, \ldots, x_m) \cdot K_T(x_1, \ldots, x_m)$.) The logical operations of 'or,' 'not,' and 'implies,' applied to primitive recursive relations, also provide primitive recursive relations.

The following three propositions illustrate how these abbreviatory practices fit together, supporting and elaborating their joint use.[3]

(1) If T is an (m + 1)-place primitive recursive relation, then so is '$\exists y \leqslant z\, T(x_1, \ldots, x_m, y)$' where $(a_1, \ldots, a_m, b) \in$ '$\exists y \leqslant z\, T(a_1, \ldots, a_m, y)$' if and only if $\exists y \leqslant b\, T(a_1, \ldots, a_m, y)$. (Proof omitted.)

(2) If $g_1, \ldots, g_v$ and h are m-place primitive recursive functions and $T_1, \ldots, T_v$ are disjoint m-place primitive recursive relations, then the function defined by

$$f(x_1, \ldots, x_m) = \begin{cases} g_1(x_1, \ldots, x_m) \text{ if } T_1(x_1, \ldots, x_m) \\ \vdots \\ g_v(x_1, \ldots, x_m) \text{ if } T_v(x_1, \ldots, x_m) \\ h(x_1, \ldots, x_m) \text{ otherwise} \end{cases}$$

is also primitive recursive. Proof: $f(x_1, \ldots, x_m) =$
$K_{T_1}(x_1, \ldots, x_m) \cdot g_1(x_1, \ldots, x_m) + \ldots + K_{T_v}(x_1, \ldots, x_m) \cdot g_v(x_1, \ldots, x_m) + (1 \dotdiv (K_{T_1}(x_1, \ldots, x_m) + \ldots + K_{T_v}(x_1, \ldots, x_m))) \cdot h(x_1, \ldots, x_m)$.

(3) Define $\mu y \leqslant z\, T(x_1, \ldots, x_m, y)$ as the least $y \leqslant z$ for which $T(x_1, \ldots, x_m, y)$ holds if there exists such a y and if not, define

PRIMITIVE RECURSIVE FUNCTIONS AND RELATIONS

$\mu y \leqslant z \ T(x_1, \ldots, x_m, y) = 0$. If T is an $(m+1)$-place primitive recursive relation, then the function defined by $(x_1, \ldots, x_m, z) \mapsto \mu y \leqslant z \ T(x_1, \ldots, x_m, y)$ is primitive recursive. Proof:

(*) $\quad \mu y \leqslant 0 \ T(x_1, \ldots, x_m, y) = 0$

$\mu y \leqslant S(z) \ T(x_1, \ldots, x_m, y) = \mu y \leqslant z \ T(x_1, \ldots, x_m, y)$ if

$\exists y \leqslant z \ T(x_1, \ldots, x_m, y)$

$= z + 1$ if $T(x_1, \ldots, x_m, S(z))$ and

$\sim (\exists y \leqslant z \ T(x_1, \ldots, x_m, y))$

$= 0$ otherwise.

(Partial exegesis: First note that $(x_1, \ldots, x_m, z) \in \ '\exists y \leqslant w \ T(x_1, \ldots, x_m, y)'$ has been replaced by $\exists y \leqslant z \ T(x_1, \ldots, x_m, y)$ and $\sim(\exists y \leqslant z \ T(x_1, \ldots, x_m, y))$ has replaced $(x_1, \ldots, x_m, z) \in$ not-'$\exists y \leqslant w \ T(x_1, \ldots, x_m, y)$.' The proof is by recursion; (*) evaluates the function at $(x_1, \ldots, x_m, 0)$, and the right side of the remaining equation defines a primitive recursive function by cases, following from Proposition (2) above.)

The last example that I will give is the proof that the relation term(x) is primitive recursive, where term(x) denotes the set of numbers corresponding under the Gödel numbering to the terms of P. As with the examples just given, this example illustrates how the various abbreviatory practices are used in conjunction with each other. However, there is an important difference between the previous examples and that of term(x). The demonstrations that numerical functions like addition, multiplication, and exponentiation are primitive recursive functions and that numerical relations like 'x divides y'[4] and 'x is a prime number' are primitive recursive relations *can be*[5] understood as definitions of addition, multiplication, etc. In contrast, the syntax of P is specified independently of any consideration of its eventual translation into numerical functions and relations via a Gödel numbering. Consequently, term(x) is, a priori, just a collection of numbers obtained through the correspondence of Lg(P)[6] to N under a specific Gödel numbering, and the demonstration that term(x) is primitive recursive is the demonstration that the characteristic function of this (a priori) 'unstructured'· set of numbers is a primitive recursive function.

In order to present such a demonstration, a precise definition of the terms of P must be given. The *terms* of P are defined inductively as only those expressions of P determined by the following conditions: (1) 0 is a term; (2) individual variables (i.e., $x_0, x_1, x_2, \ldots$) are terms; (3) if v is a term, then so is $S(v)$; and (4) if v_1 and v_2 are terms, then $+(v_1 \ v_2)$ and $\cdot(v_1 \ v_2)$ are terms, also.[7] A term can be thought of as potentially naming a natural number in P; $S(0)$ 'names' 1 in P, $S(v)$

61

PRIMITIVE RECURSIVE FUNCTIONS AND RELATIONS

'names' the successor of v. The other definitions that will be needed are those of L(x), $(x)_i$, x * y, and var(x). L(x) counts the number of primes in the prime factorization of x; $(x)_i$ gives the exponent of the i-th prime in the prime factorization of x; and x * y is the function that 'concatenates' the prime factorizations of x and y. All three of these functions are primitive recursive. var(x) is the primitive recursive relation that holds if and only if x is the Gödel number of an individual variable of P.

A materially adequate demonstration that the set of Gödel numbers corresponding to the terms of P is a primitive recursive relation can now be given, and it consists of the formula[8]

$$\text{term}(x) \Leftrightarrow x \neq 0 \text{ and } \exists y \leqslant \prod_{i=0}^{L(x)} p_i^x \Big\{ [(y)_{L(y)} = x] \text{ and}$$
$$\forall i \leqslant L(y) \,([i = 0] \text{ or } [(y)_i = g(0)] \text{ or}$$
$$[\text{var}((y)_i)] \text{ or } \exists j < i \,[(y)_i = g(S(\ ) * (y)_j * g(\ ))]$$
$$\text{or } \exists j < i \,\exists k < i \,([(y)_i = g(+(\ ) * (y)_j * (y)_k * g(\ ))]$$
$$\text{or } [(y)_i = g(\cdot(\ ) * (y)_j * (y)_k * g(\ ))]))\Big\}.$$

Given its placement in the course of work that makes up the proving of Gödel's theorem, this formula seeably/showably defines a primitive recursive relation. In that the formula makes available a methodic procedure for assigning to each and only the terms of P a number that satisfies it, the adequacy of the formula as exactly defining the Gödel numbers of the terms can be checked by 'reasoned' calculations comparing the formula to that numbering and to various examples of terms constructed as 'typical' cases. Finally, that the bound on y (i.e., $\prod_{i=0}^{L(x)} p_i^x$ is large enough is established through considerations of the structure of the Gödel numbering as that structure becomes available over the course of developing both the formula and the bound. Thus, together, these observations point to the in-situ determination of the adequacy of the formula defining term(x) as a primitive recursive relation as being tied to the local practices that make up the work of writing and inspecting that formula.

With these examples before us, it is possible to clarify certain aspects of the abbreviatory practices used in working with primitive recursive functions and relations. By doing so, we will be led to the pointed relevance of these practices for the proof of Gödel's theorem.

1 The use and omnipresence of these practices are almost transparent. The abbreviatory practices are introduced gradually, coincident with the increasingly extended constructions of primitive recursive functions and relations, and when they are introduced, they have the

PRIMITIVE RECURSIVE FUNCTIONS AND RELATIONS

initial character of being either notational conventions or theorems of ordinary mathematics. The cumulative effect of these practices as a body of learned, practical techniques for working with primitive recursive functions and relations makes its appearance only when the presentation of a proof of primitive recursiveness — like that of term(x) above — is disengaged from the sequence of such propositions and proofs that was organized to include that proof at its particular place in that sequence. For the reader of the proof of Gödel's theorem, the proofs of the series of propositions and proofs concerning the construction of the necessary primitive recursive functions and relations are understood *practically* not as self-suffcent demonstrations, but as guides and summaries of the work that, in the end, finds the adequacy of them. The work of finding the practical accountability of that sequence of propositions and proofs serves to instruct the reader in the abbreviatory practices that are being used. As we shall see, for the person engaged in proving Gödel's theorem, the abbreviatory practices are essential to constructing the practically objective proof that a function or relation is primitive recursive. In either case, the transparency of those practices turns out to consist of the work of producing the practically objective proofs of that primitive recursiveness.

2 From within the work of their proofs, the informal demonstrations that certain functions and relations are primitive recursive are not viewed as standing proxy for a formal rendering of them; they are the practically adequate demonstrations of the propositions for which they stand as proofs. That they can, later, be considered as representing formal proofs is essentially tied to the ways in which the abbreviatory practices are developed over the course of increasingly extended constructions of primitive recursive functions and relations and to the ways in which the informal proofs can be given increasing detail as approximations of formal ones.

3 Although introduced and reflectively considered as 'abbreviatory practices,' these practices, in fact, make up the learned, practical techniques of working with primitive recursive functions and relations. The point is not that formal proofs cannot be given nor that these practices are not adequate to the construction of such formal proofs. The point is that the practical objectivity and accountability of the informal demonstrations does not depend on giving such formal proofs, but is available in and as the practical techniques of working with primitive recursive functions and relations.

The existence of these abbreviatory practices and their connection with the practical techniques of working with primitive recursive functions and relations leads to the major point of the immediate discussion. *The real problem in working with primitive recursive functions and relations, at least for the proof of Gödel's theorem, is that of accountably showing that each of a sequentialized series of functions and relations is, in fact, primitive recursive.* The way that this is done is (1) by writing — where 'writing' is a trivialization of the work of writing at the proper place in a sequentialized series of such writings — a practically objective formula or formulas for a function or relation in terms of known primitive recursive functions and relations and in terms of known ways of building primitive recursive functions and relations that seeably/showably insures the primitive recursiveness of it, and (2) by checking in a locally determined, practically adequate manner that the formula or formulas do compute the correct values, where the calculations themselves are 'reasoned' procedures, practically adequate to the task at hand.

Consider for a moment a formal construction sequence for ordinary multiplication of natural numbers — $I_1^1, I_3^3, S, S(I_3^3)$,

$$\left\{ \begin{array}{l} +(x, 0) = I_1^1(x) \\ +(x, S(y)) = (S(I_3^3))\,(x, y, +(x, y)) \end{array} \right\}, I_1^3, +(I_1^3, I_3^3), Z,$$

$$\left\{ \begin{array}{l} \cdot(x, 0) = Z(x) \\ \cdot(x, S(y)) = (+(I_1^3, I_3^3))\,(x, y, \cdot(x, y)) \end{array} \right\} \text{ — and compare it with}$$

the 'abbreviated proof'

$x \cdot 0 = 0$

$x \cdot S(y) = xy + x.$

The formal primitive recursive equations do not make apparent what ordinary numerical function they define, and it would be difficult to find the formal primitive recursive equations without having recourse to the abbreviated proof. The difficulty that is illustrated here in the case of multiplication becomes severe as the functions and relations involved become more complicated. In that the adequacy of an equation defining a primitive recursive function or relation as a description of a pre-given function or relation is made available through the work of writing and inspecting that formula as seeably/showably adequate to that claim, the point to be made is this: *the abbreviatory practices/ practical techniques make that writing and inspection possible.* It is this aspect of the abbreviatory practices that is crucial to the practical objectivity — to the 'rigor' — of the work of proving Gödel's theorem.

7 A Schedule of Proofs:
An Extended Analysis of the Lived-Work of Producing the Body of a Proof of Gödel's Theorem

A
A Schedule of Proofs

The body of the proof of Gödel's theorem consists of a schedule of proofs that certain functions and relations are primitive recursive and that certain ways of building functions and relations from other functions and relations preserve primitive recursiveness, that schedule leading to, and organized so as to permit, the construction of G in the diagonalization/'proof' as a primitive recursive relation. The reader will recall that the wff $G(x_1, x_2)$ numeralwise expressing G in P was used to construct the formally undecidable sentence J and that the numeralwise expressibility of G was used in both parts of the 'proof' to transfer facts about membership in G (e.g., G(n, i)) to *theorems* of P (e.g., $\vdash_P G(k(n), k(i))$). Once the schedule of proofs has been completed and the theorem asserting the numeralwise expressibility of primitive recursive relations in P has been established, the diagonalization/'proof' can be constructed, bringing to an end the proof of Gödel's theorem.

In this section, I recall for the reader the general pattern of ordering of such a schedule of proofs[1] and then describe some of the general features of that schedule of proofs of the lived work of its production. By a 'schedule of proofs,' I refer both to the ordered sequence of proportions of that schedule and to the proofs of those propositions, the latter generally consisting of a single formula. I have omitted these proofs from the outline that follows. A few marginal comments have been added to indicate the schedule's topical organization.

1 The constant functions $Z_n(x) = n$, n = 0, 1, 2, ... are primitive recursive.

2 The functions obtained by permuting variables, identifying variables, adding dummy variables, and substituting constants for variables in

A SCHEDULE OF PROOFS

primitive recursive functions are primitive recursive.

3 Addition, multiplication, exponentiation, and 'limited subtraction' are primitive recursive functions.

4 Let f be an $(m + 1)$-place numerical function and (informally) let $\sum_{k=0}^{y} f(x_1, \ldots, x_m, k)$ and $\prod_{k=0}^{y} f(x_1, \ldots, x_m, k)$ denote the functions $(x_1, \ldots, x_m) \mapsto f(x_1, \ldots, x_m, 0) + \ldots + f(x_1, \ldots, x_m, y)$ and $(x_1, \ldots, x_m, y) \mapsto f(x_1, \ldots, x_m, 0) \cdot \ldots \cdot f(x_1, \ldots, x_m, y)$, respectively. Then if f is a primitive recursive function, so are $\sum_{k=0}^{y} f(x_1, \ldots, x_m, k)$ and $\prod_{k=0}^{y} f(x_1, \ldots, x_m, k)$.

5 $=, \neq, <$, and $\leq$ are primitive recursive relations.

6 The logical operations of 'not,' 'and,' 'or,' 'implies,' and 'if and only if,' applied to primitive recursive relations, produce primitive recursive relations.

7 The relation obtained by substituting a primitive recursive function for a variable in a primitive recursive relation is primitive recursive.

8 Let R be an $(m + 1)$-place numerical relation and let $\forall z \leq y\ R(x_1, \ldots, x_m, z)$ and $\exists z \leq y\ R(x_1, \ldots, x_m, z)$ denote the $(m + 1)$-place relations that hold for $(x_1, \ldots, x_m, y)$ if $R(x_1, \ldots, x_m, z)$ holds for all $z \leq y$ or for some $z \leq y$, respectively. If R is a primitive recursive relation, then so are $\forall z \leq y\ R(x_1, \ldots, x_m, z)$ and $\exists z \leq y\ R(x_1, \ldots, x_m, z)$.

9 If $T_1, \ldots, T_v$ are pairwise disjoint, m-place primitive recursive relations, and if $g_1, \ldots, g_v$ and h are m-place primitive recursive functions, then the function f defined by the equation

$$f(x_1, \ldots, x_m) = \begin{cases} g_1(x_1, \ldots, x_m) & \text{if } T_1(x_1, \ldots, x_m) \\ g_2(x_1, \ldots, x_m) & \text{if } T_2(x_1, \ldots, x_m) \\ \quad \vdots \\ g_v(x_1, \ldots, x_m) & \text{if } T_v(x_1, \ldots, x_m) \\ h(x_1, \ldots, x_m) & \text{if } (x_1, \ldots, x_m) \notin T_i \\ & \text{for all } i, 1 \leq i \leq v. \end{cases}$$

is primitive recursive.

10 If R is an $(m + 1)$-place numerical relation, let $\mu z \leq y\ R(x_1, \ldots, x_m, z)$ (informally) denote the function of $(x_1, \ldots, x_m, y)$ defined by the equation

A SCHEDULE OF PROOFS

$$\mu z \leqslant y\ R(x_1, \ldots, x_m, z) = \begin{cases} \text{the least } z \leqslant y \text{ such that } R(x_1, \ldots, x_m, z) \text{ if there is such a } z \\ 0 \quad \text{otherwise.} \end{cases}$$

If R is an (m + 1)-place primitive recursive relation, then $\mu z \leqslant y\ R(x_1, \ldots, x_m, z)$ is a primitive recursive function.

The preceding propositions provide the background facts and techniques that are needed for the proofs of the propositions that follow. The next group of propositions supply the apparatus for working with the Gödel numbering that was introduced earlier.

11 The 2-place numerical relation $x|y$ ('x divides y with no remainder') is primitive recursive.

12 Let prime(x) hold if and only if x is a prime number. Then prime(x) is a primitive recursive relation.

13 The function p_n (i.e., $n \mapsto p_n$) giving, for each n, the n-th prime number is primitive recursive ($p_0 := 1$).

14 The 'decoding function' $(x)_n$ (i.e., $(x, n) \mapsto (x)_n$) giving the exponent of p_n in the prime factorization of x is a primitive recursive function.[2]

$((x)_0 := 0; (0)_n := 0.)$

15 Let the function L(x) give the number (in the serial ordering of the prime numbers) of the largest prime with a non-zero exponent in the prime factorization of x or give 0 if x is 0 or 1.[3] L(x) is a primitive recursive function.

16 For every natural number y, $y = 0$ or $y = p_0^{(y)_0} p_1^{(y)_1} \cdot \ldots \cdot p_{L(y)}^{(y)_{L(y)}}$. Define (informally) x * y as the function mapping (x, y) to the value

$$x * y = x \cdot \prod_{k=0}^{L(y)} p_{L(x)+k}^{(y)_k} \ .$$

x * y is a primitive recursive function.

The remaining propositions assert the primitive recursiveness of the functions and relations that correspond under the previously specified Gödel numbering to certain syntactic categories and operations of the formal system P.

17 The mapping of a natural number n to the Gödel number of the individual variable x_n (i.e., $n \mapsto g(x_n)$) is a primitive recursive function.

18 Let var(x) be the relation that holds if and only if x is the Gödel number of a variable of P. var(x) is primitive recursive.

19 formterm(x), holding if and only if x is the Gödel number of a

A SCHEDULE OF PROOFS

'formation sequence of terms,' is a primitive recursive relation.
20 Define the relation term(x) as holding if and only if x is the Gödel number of a term of P. term(x) is primitive recursive.
21 The relation formwff(x), holds exactly when x is the Gödel number of a 'formation sequence of wffs,' is primitive recursive.
22 wff(x) is a primitive recursive relation, where wff(x) holds if and only if x is the Gödel number of a wff of P.
23 Let occur(w, x, y, z) hold exactly when z is the Gödel number of a wff, x is the Gödel number of a variable, and z can be written as z = w * x * y.[4] occur(w, x, y, z) is a primitive recursive relation.
24 bound(w, x, y, z), indicating roughly that an occurrence of the variable with Gödel number x is bound in the wff with Gödel number z, is a primitive recursive relation.
25 The relations free(w, x, y, z), which holds if an only if occur(w, x, y, z) and not-bound(w, x, y, z), is primitive recursive.
26 Let S(x, t, a) give the Gödel number of the wff that results when one free occurrence of the variable with Gödel number x in the wff with Gödel number a is replaced by the term with Gödel number t if x is the Gödel number of a variable x, t is the Gödel number of a term, a is the Gödel number of a wff A, and x does occur free in A, and let S(x, t, a) = a otherwise. S(x, t, a) is a primitive recursive function.
27 S^m(x, t, a), iterating the operation of S(x, t, a) m times, is a primitive recursive function of m, x, t, and a.
28 The replacement, conveyed through Gödel numbers, of the variable x throughout a wff A by a formulaically determined variable occurring neither in A nor in a specified term t is a primitive recursive function of x, t, and a where x, t, and a are the Gödel numbers corresponding to x, t, and A, respectively.
29 Define Sub(x, t, a) as giving the Gödel number of the wff that results when the term with Gödel number t is substituted for the free occurrences of the individual variable with Gödel number x in the wff with Gödel number a when, in fact, x is the Gödel number of an individual variable, t is the Gödel number of a term, and a is the Gödel number of a wff, and as giving a otherwise.[5] Sub(x, t, a) is a primitive recursive function.
30 Let sub(x, n, a) equal the Gödel number of the wff that results when k(n) is substituted for the free occurrences of the individual variable with Gödel number x in the wff with Gödel number a when, in fact, x is the Gödel number of an individual variable and a is the Gödel number of a wff, and let it equal a otherwise.[6] sub(x, t, a) is a primitive recursive function.
31 freefor(x, t, a) is a primitive recursive relation, where freefor(x, t, a) holds if and only if x, t, and a are the Gödel numbers of a variable x, a term t, and a wff A, respectively, and if 't is free for x in A.'

A SCHEDULE OF PROOFS

32 The relation notfree(x, a), which holds if and only if x is the Gödel number of a variable with no free occurrences in a wff with Gödel number a, is a primitive recursive relation.

33 $axiom_1(x), \ldots, axiom_i(x), \ldots, axiom_k(x)$ are primitive recursive relations, where i enumerates the axioms of P and $axiom_i(x)$ holds if and only if x is the Gödel number of an instance of axiom i.

34 Define the relation deduct(x) as holding if and only if x is the Gödel number of a deduction. deduct(x) is primitive recursive.

35 The relation ded(x, y), which holds only when x is the Gödel number of a deduction of the wff with Gödel number y, is a primitive recursive relation.

Proposition 35 and its proof complete a schedule of proofs for the proof of Gödel's theorem. Once sub(x, n, a) and ded(x, y) have been shown to be primitive recursive, the practical techniques of working with primitive recursive functions and relations make available (and are available as) the seeable/showable adequacy of the demonstration that G(x, u) is a primitive recursive relation:

$G(x, u) \Leftrightarrow ded(x, \phi(u))$

$\Leftrightarrow ded(x, sub_{g(x_2)}(u, u))$

$\Leftrightarrow ded(x, sub(g(x_1), u, u))$.

Thus, once the lemma asserting the numeralwise expressibility of primitive recursive relations in P has been established, the diagonalization/ 'proof' can be constructed and the proof of Gödel's theorem completed.

B
A Schedule of Proofs as Lived-Work

In the remainder of this chapter, I describe four general features of the schedule of proofs in terms of the lived-work of producing that schedule. The material that is introduced here prepares the reader for the analysis in the next chapter of what identifies a schedule of proofs as a naturally accountable schedule of proofs for a proof of Gödel's theorem.

1 *As it is used in developing the schedule of proofs, a Gödel numbering is not an abstractly defined correspondence between the symbols, expressions, and sequences of expressions of P and the natural numbers; it is a technique of proving*

Disengaged from the work of proving Gödel's theorem, a Gödel numbering merely establishes a correspondence between the symbols, expressions and sequences of expressions of P and the natural numbers.

A SCHEDULE OF PROOFS

Under this correspondence, the syntax of P can be translated into numerical functions and relations, but, speaking again in a manner disengaged from the work of proving Gödel's theorem, the numerical functions and relations so defined are in no way assured analyzable mathematical structures. In practice, however, a specific Gödel numbering does provide for the mathematical analyzability of these functions and relations.

As a hypothetical example, suppose that the numerical relation R corresponding to some syntactic feature $\mathcal{R}$ of P under the Gödel numbering with which we are working consists of all the natural numbers for which the exponent of the largest prime in each of these numbers' prime factorizations is 2 — that is, R is the set of numbers 2^2, $2 \cdot 3^2$, $2^2 \cdot 3^2$, $2 \cdot 3 \cdot 5^2$, $2^2 \cdot 3^2 \cdot 5^2$, $2^3 \cdot 3 \cdot 5^2$, etc. Given the apparatus for working with primitive recursive functions and relations developed in Propositions 1 through 16, the definition of R can be written as

$$R(x) \Leftrightarrow (x)_{L(x)} = 2.$$

The immediate point is that this formula, besides being the seeable/showable counterpart of the verbal definition of R, seeably/showably exhibits R as a primitive recursive relation.

The artificiality of this example comes from the fact that $\mathcal{R}$ is presented in terms of the structure of the numerical relation R corresponding to it.[7] In contrast, in that the syntax of P is defined prior to, and independently of, the specification of a Gödel numbering, the functions and relations appearing in Propositions 19 through 35 do not have a predetermined mathematical structure. 'Gödel numbering as a technique of proving' refers to the fact that a proof-specific Gödel numbering not only allows the prover to define these numerical functions and relations, but allows him to demonstrate, through the work of uncovering the analyzable mathematical structure that that numbering provides, that these functions and relations are primitive recursive. The relation term(x), for example, is defined only as the image under the Gödel numbering of the terms of P. It is the technique of working with that Gödel numbering that makes this set of numbers accessible to the proof that it is, in fact, a primitive recursive relation.

As a means of recalling for the reader this technique of proving, I conclude the discussion of it by outlining the work of producing a formula that defines term(x) as a primitive recursive relation.

The reader will recall that the terms of P are defined inductively and include only those expressions of P determined by the following conditions: (1) 0 is a term, (2) the individual variables are terms, (3) if ν is a term, then so is $S(\nu)$, (4) if ν_1 and ν_2 are terms, then so is $+(\nu_1 \nu_2)$, and (5) if ν_1 and ν_2 are terms, then so is $\cdot(\nu_1 \nu_2)$. term(x) is then defined as the image of the terms of P under the Gödel numbering, or, in other words, term(x) holds if and only if x is the Gödel number of

A SCHEDULE OF PROOFS

an expression of P satisfying one of the five conditions just listed. The proof that term(x) is primitive recursive consists of translating this definition into a mathematical formula that (i) given its placement in the schedule of proofs, seeably/showably exhibits the primitive recursiveness of the relation that it defines, and, simultaneously, (ii) can be read to find that the relation defined by the formula exactly identifies the Gödel numbers of the terms of P.

Let us begin the work of proving that term(x) is primitive recursive by examining the case that x is the Gödel number of 0. A numerical relation consisting of a single number a is primitive recursive, as can be justified by the fact that, since = is a primitive recursive relation, Propositions 2 and 7 imply that $\{x \mid x = a\}$ is primitive recursive. In that we are only interested in the defining condition $x = g(0)$, we may write

term(x) $\Leftrightarrow$ $x = g(0)$ or . . . ,

and, by Proposition 6, if the condition following 'or' defines a primitive recursive relation, term(x) will be primitive recursive, also.

Next, consider the case, denoted var(x), that x is the Gödel number of an individual variable of P. As before, let 3, 5, 7, and 9 be the Gödel numbers assigned to the primitive symbols (,), ', and x, respectively. If the individual variables of P are enumerated as follows

$x_0 := (x)$

$x_1 := (x')$

$x_2 := (x'')$

.
.
$x_n := (x''\overbrace{}^{n}\,')$
.
.

x will be the Gödel number of an individual variable if and only if $x \in \{g(x_0), g(x_1), g(x_2), \ldots, g(x_n), \ldots\}$. Computing the values for several of the $g(x_j)$, beginning with $i = 0$, one obtains the collection $\{2^3 \cdot 3^9 \cdot 5^5,\ 2^3 \cdot 3^9 \cdot 5^7 \cdot 7^5,\ 2^3 \cdot 3^9 \cdot 5^7 \cdot 7^7 \cdot 11^5,\ 2^3 \cdot 3^9 \cdot 5^7 \cdot 7^7 \cdot 11^7 \cdot 13^5,\ \ldots\}$. The orderly way in which these numbers are arrived at points to the formula for $g(x_n)$ as $2^3 \cdot 3^9 \cdot 5^7 \cdot \ldots \cdot p_{n+2}^7 \cdot p_{n+3}^5$. With the apparatus of Propositions 13 through 16 as background, this may be written as the seeably/showably primitive recursive function

$$n \mapsto g(x_n) = 2^3 \cdot 3^9 * \prod_{i=0}^{n} p_i^7 * 2^5 = g((x) * \prod_{i=0}^{n} p_i^7 * g(\ )).$$

A SCHEDULE OF PROOFS

Let us return to the problem of demonstrating that the relation 'x is the Gödel number of an individual variable' is primitive recursive. The condition that x is the Gödel number of an individual variable translates into 'x = $g(x_n)$ for some n,' and, by writing this more formally, the formula

$$\text{var}(x) \Leftrightarrow \exists n \, (x = g(x_n))$$

is obtained. Although *this* formula does not exhibit var(x) as a primitive recursive relation, the practical techniques of working with primitive recursive functions and relations insure the primitive recursiveness of var(x) if (i) the relation x = $g(x_n)$ can be shown to be primitive recursive and (ii) an upper bound for n as a function of x alone can be found. But $g(x_n)$ was just shown to be a primitive recursive function, implying that x = $g(x_n)$ defines a primitive recursive relation. Since n will always be less than $g(x_n)$,[8] we have

$$\text{var}(x) \Leftrightarrow \exists n \leq x \, (x = g(x_n))$$

seeably/showably defining var(x) as a primitive recursive relation,[9] and

$$\text{term}(x) \Leftrightarrow x = g(0) \text{ or var}(x) \text{ or } \ldots$$

will define a primitive recursive relation if the condition following the second 'or' defines one.

Consider, now, the remaining three cases — that x is the Gödel number of an expression of P either of the form (3) $S(v)$, (4) $+(v_1 v_2)$, or (5) $\cdot(v_1 v_2)$, where v, v_1 and v_2 are themselves terms. By temporarily using 'corners' to indicate the Gödel number of a primitive symbol, these conditions can be written as

(3a) term(v) and x = $2^{\ulcorner S \urcorner} \cdot 3^{\ulcorner (\urcorner * v * 2^{\ulcorner) \urcorner}}$

(4a) term(v_1) and term(v_2) and x = $2^{\ulcorner + \urcorner} \cdot 3^{\ulcorner (\urcorner * v_1 * v_2 * 2^{\ulcorner) \urcorner}}$

(5a) term(v_1) and term(v_2) and x = $2^{\ulcorner \cdot \urcorner} \cdot 3^{\ulcorner (\urcorner * v_1 * v_2 * 2^{\ulcorner) \urcorner}}$

where term(v), of course, indicates that v is the Gödel number of a term. Without using 'corners,' these conditions become

(3b) term(v) and x = g(S( ) * v * g( ))

(4b) term(v_1) and term(v_2) and x = g(+( ) * v_1 * v_2 * g( ))

(5b) term(v_1) and term(v_2) and x = g(·( ) * v_1 * v_2 * g( ))

g being used exclusively for the Gödel numbers of expressions. In that the right side of each of the equations for x seeably/showably defines a primitive recursive function of either v or of both v_1 and v_2, the equations for x define primitive recursive relations of x and, appropriately, of either v or of v_1 and v_2.

Let us examine the case that x is the Gödel number of a term of the

A SCHEDULE OF PROOFS

form $S(v)$. Someone engaged in proving Gödel's theorem might begin developing a formula that exhibits the primitive recursiveness of this relation by first translating it into the formalism

$\exists v \, [\text{term}(v) \text{ and } x = g(S(\ ) * v * g(\ ))]$

and then proceed by finding a bound on v depending on x alone. In that consideration of the structure of the Gödel numbering shows that the Gödel number of an expression will always be greater than the Gödel number of a 'part' of it,[10] x itself will be a bound on v. Thus, the prover arrives at the formula

$\exists v \leqslant x \, [\text{term}(v) \text{ and } x = g(S(\ ) * v * g(\ ))]$.

However, the real problem remains: the apparent circularity of this formula in a definition of term(x) must be removed and it must be removed in such a way that the formula that results defines term(x) as a primitive recursive relation.

By proceeding in a similar manner for the cases of $+(v_1 v_2)$ and $\cdot(v_1 v_2)$, one obtains the formulas

(4c) $\exists v_1 \leqslant x \, \exists v_2 \leqslant x \, [\text{term}(v_1) \text{ and } \text{term}(v_2) \text{ and }$

$x = g(+(\ ) * v_1 * v_2 * g(\ ))]$

and

(5c) $\exists v_1 \leqslant x \, \exists v_2 \leqslant x \, [\text{term}(v_1) \text{ and } \text{term}(v_2) \text{ and }$

$x = g(\cdot(\ ) * v_1 * v_2 * g(\ ))]$,

again presenting the problem of their impredicativity but with the further complication that two possibly distinct variables are being specified as the Gödel numbers of terms.

Before I give the construction of a formula seeably/showably exhibiting term(x) as a primitive recursive relation, an important point must be made. In using a given solution to find how the Gödel numbering provides for a numbering of the terms of P adequate to that solution and, at the same time, in using that solution to find the method of constructing the terms of P that makes that numbering possible, *the reader* works toward the recovery of pre-existent, objective structures of the syntax of P and of the Gödel numbering. In contrast, for someone engaged in developing the schedule of proofs, the problem of removing the impredicativity in the definition of term(x) and cotemporaneously showng that term(x) is primitive recursive is the problem of finding a method, using the Gödel numbering, of associating numbers with the terms of P with the property that the structure of that method, as it is made available through the use of the apparatus for working with the Gödel numbering, identifies exactly the Gödel numbers of the terms. In other words, *the prover*

A SCHEDULE OF PROOFS

looks to discover that aspect of the structure of the techniques of working with the Gödel numbering that will provide for what a solution could consist of. For the prover, the solution is not molded to the known structure of the Gödel numbering and the definition of the terms of P; in the presence of the problem, the prover uses and develops the techniques of working with the Gödel numbering to discover a mathematically identifiable method of numbering the terms adequate to what a solution then becomes.

As a way of working toward the discovery of a formula for term(x), consider a particular term t of P, for example,[11]

$$t = +(x_1 S(\cdot(x_5 0))).$$

t can be viewed as being built up in a succession of steps, each in accord with the inductive definition of the terms of P:

$$x_5, 0, \cdot(x_5 0), S(\cdot(x_5 0)), x_1, +(x_1 S(\cdot(x_5 0))).$$

Denoting this sequence of terms as τ, τ will be referred to as a formation sequence for t. The point of constructing such a formation sequence is that a Gödel number y can be assigned to it, namely

$$y = p_1^{g(x_5)} \cdot p_2^{g(0)} \cdot p_3^{g(\cdot(x_5 0))} \cdot p_4^{g(S(\cdot(x_5 0)))} \cdot p_5^{g(x_1)} \cdot p_6^{g(+(x_1 S(\cdot(x_5 0))))},$$

with the following properties: (i) the apparatus of Propositions 13 through 16 allows each of the exponents of the prime factors of y to be checked one at a time to see if it is the Gödel number of a term — provided there is some way to make that determination, and (ii) given that y represents the formation sequence of a *particular* term t, it is not necessary to examine whether or not $g(t) = g(S(v))$ or $g(t) = g(+(v_1 v_2))$ or $g(t) = g(\cdot(v_1 v_2))$ for all possible terms v or v_1 and v_2, but only whether or not one of these equations holds for terms v or v_1 and v_2 with Gödel numbers occurring as the exponents of primes smaller than $p_{L(y)}$ in the prime factorization of y. In fact, (ii) holds for the exponent of any prime factor of y when one replaces $p_{L(y)}$ with the prime number of which it is the exponent.

The generalization of the idea of this construction and numbering proceeds by mathematically identifying that construction and numbering as a methodic procedure: first, a *formation sequence of terms* is defined as a finite sequence of expressions of P — $t_1, \ldots, t_i, \ldots, t_m$ — such that, for each i, i = 1, ..., m, one of the following conditions holds:

(i) t_i is 0
(ii) t_i is an individual variable
(iii) t_i is $S(t_j)$ for some j < i
(iv) t_i is $+(t_j t_k)$ for some j < i, k < i
(v) t_i is $\cdot(t_j t_k)$ for some j < i, k < i.

A SCHEDULE OF PROOFS

The terms of P can then be seen to be only those expressions of P that are the final expressions in a formation sequence of terms.

Let t be a term and let τ be a formation sequence for it:

(6) $\tau = t_1, \ldots, t_j, \ldots, t_k, \ldots, t_i, \ldots, t_m$,

where $t_m = t$. The Gödel number of τ will be

(7) $y = p_1^{g(t_1)} \cdot \ldots \cdot p_j^{g(t_j)} \cdot \ldots \cdot p_k^{g(t_k)} \cdot \ldots \cdot p_i^{g(t_i)} \cdot \ldots \cdot p_m^{g(t_m)}$.

Given the availability of this numbering, a number x will be the Gödel number of a term if and only if there exists some number y such that the last exponent in the prime factorization of y is x (i.e., $(y)_{L(y)} = x$) and, for each exponent in the prime factorization of v (i.e., $\forall_i \leq L(y)$, $i \neq 0$), that exponent is either the Gödel number of 0 (i.e., $(y)_i = g(0)$) or the Gödel number of a variable (i.e., $var((y)_i)$) or the Gödel number of an expression of the form $S(t_j)$ for some expression t_j where the Gödel number of t_j is an exponent of some p_j in the prime factorization of y with $p_j < p_i$ (i.e., $\exists j < i \, [(u)_i = g(S(\;) * (v)_j * g(\;))])$ or ...

The rendering into standard English of these conditions is both unnecessary and obfuscating. The prover's familiarity with the practices of working with primitive recursive functions and relations allows him to work directly from the method of numbering indicated in (6) and (7) and formulate the conditions on term(x) by rendering them into the apparatus for constructing primitive recursive relations. In this way, the prover directly constructs the formula

$\exists y \, \{[(y)_{L(y)} = x] \text{ and } \forall i \leq L(y) \, ([i = 0] \text{ or } [(y)_i = g(0)]$

or $[var((y)_i)]$ or $\exists j < i \, [(y)_i = g(S(\;) * (y)_j * g(\;))]$

or $\exists j < i \, \exists k < i \, ([(y)_i = g(+(\;) * (y)_j * (y)_k * g(\;))]$

or $[(y)_i = g(\cdot(\;) * (y)_j * (y)_k * g(\;))])])\}$,

seeably/showably defining a primitive recursive relation in x if a bound for y depending on x alone can be found.[12]

In order to find this bound, let us return[13] to the juxtaposition of t, τ, and y:

$t = +(x_1 S(\cdot(x_5 0)))$

$\tau = x_5, 0, \cdot(x_5 0), S(\cdot(x_5 0)), x_1, +(x_1 S(\cdot(x_5 0)))$

$y = p_1^{g(x_5)} \cdot p_2^{g(0)} \cdot p_3^{g(\cdot(x_5 0))} \cdot p_4^{g(S(\cdot(x_5 0)))} \cdot p_5^{g(x_1)}$
$\cdot p_6^{g(+(x_1 S(\cdot(x_5 0))))}$.

Here, one notices that at least one formation sequence for t (i.e., τ) will have six constituents, one for each primitive symbol of t that is not a parenthesis. Thus, there will be at least one number y (i.e., the y

75

A SCHEDULE OF PROOFS

computed above) such that y is the Gödel number of a formation sequence of t and y has the first six prime numbers as its only factors. Let $x = g(t)$. In that the function $L(x)$ counts the parentheses as well as the other symbols in t, $L(y)$ will be less than or equal to $L(x)$. Further, in that the Gödel number of a 'part' of an expression is less than the Gödel number of the expression itself,[14]

$$y = p_1^{g(x_s)} \cdot p_2^{g(0)} \cdot p_3^{g(\cdot(x_s,0))} \cdot p_4^{g(S(\cdot(x_s,0)))} \cdot p_5^{g(x_1)}$$
$$\cdot p_6^{g(+(x_1(\cdot(x_s,0))))}$$

$$\leq p_1^x \cdot \ldots \cdot p_6^x \leq p_1^x \cdot \ldots \cdot p_{L(x)}^x = \prod_{i=0}^{L(x)} p_i^x,$$

and $\prod_{i=0}^{L(x)} p_i^x$ will be a bound on y.

The methodic character of this computational reasoning in the case of the immediate example provides the practical accountability of its generalization. Thus, if there is no $y \leq \prod_{i=0}^{L(x)} p_i^x$ such that y is the Gödel number of a formation sequence of a term with Gödel number x, x will not be the Gödel number of a term.

Finally, a review of the original formula to examine its adequacy shows that it excludes the case that $x = 1$, but not that of $x = 0$, both 0 and 1 not being the Gödel numbers of terms.[15] Adding this condition, one obtains the formula

$$\text{term}(x) \Leftrightarrow x \neq 0 \text{ and } \exists y \leq \prod_{i=0}^{L(x)} p_i^x \{[(y)_{L(y)} = x] \text{ and }$$

$$\forall i \leq L(y) ([i = 0] \text{ or } [(y)_i = g(0)] \text{ or }$$

$$[\text{var}((y)_i)] \text{ or } \exists j < i [(y)_i = g(S(\ ) * (y)_j * g(\ ))]$$

$$\text{or } \exists j < i \exists k < i ([(y)_i = g(+(\ ) * (y)_j * (y)_k * g(\ ))]$$

$$\text{or } [(y)_i = g(\cdot(\ ) * (y)_j * (y)_k * g(\ ))]])\}$$

seeably/showably defining term(x) as a primitive recursive relation.

2 *The schedule of proofs has a 'directed' character: it leads to and is organized so as to permit, the construction of G as a primitive recursive relation*

An initial sense of the 'directed' character of the schedule of proofs can be obtained by working back from the definition of G and establishing 'lines of dependence' between the various propositions of that schedule.

A SCHEDULE OF PROOFS

I will briefly indicate how this can be done.

Let us begin with the construction of G. That G is primitive recursive can be seen to depend on the demonstrations that sub and ded are primitive recursive and, further, on the use of Propositions 2 and 7.[16] If we leave aside the matter of the placement of Propositions 2 and 7, the dependence of the primitive recursiveness of G on that of sub and ded can be depicted graphically as follows:

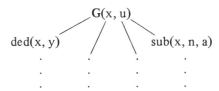

Working backwards, consider the demonstration that ded(x, y) is primitive recursive. Let $A_1, \ldots, A_m = A$ be a deduction of the wff A, $a_1, \ldots, a_m = a$ be the Gödel numbers corresponding to $A_1, \ldots, A_m$, respectively, and let $b = p_1^{a_1} \cdot \ldots \cdot p_m^{a_m}$. Then b is the Gödel number of a deduction of the wff with Gödel number a — that is, ded(b, a). But if $A'_1, \ldots, A'_s = A$ is another deduction of A and if $a'_1, \ldots, a'_s = a$ are the Gödel numbers corresponding to $A'_1, \ldots, A'_s$, then $b' = p_1^{a'_1} \cdot \ldots \cdot p_s^{a'_s}$ is also the Gödel number of a — ded(b', a). From these examples it can be seen that ded(x, y) holds if and only if x is the Gödel number of a deduction (i.e., deduct(x)) and if the exponent of the largest prime number in the prime factorization of x is the Gödel number of y. In symbols,

ded(x, y) ⇔ deduct(x) and $(x)_{L(x)} = y$.

Given its placement in the schedule of proofs, this equation constitutes the materially adequate proof that ded(x, y) is a primitive recursive relation,[17] but for now, its importance for us lies in the fact that it exhibits the dependence of the primitive recursiveness of ded(x, y) on that of earlier propositions in the schedule of proofs. Graphically, we have the following situation:

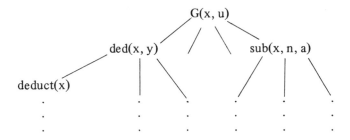

A SCHEDULE OF PROOFS

As one last example, let us examine the proof of the primitive recursiveness of the relation deduct(x) which holds exactly when x is the Gödel number of a deduction. A deduction in P is defined as a finite sequence of wffs $A_1, \ldots, A_i, \ldots, A_m$ subject to the conditions that each wff A_i of that sequence is either an instance of one of the axioms of P, A_i results from some A_s, $s < i$, by generalization — that is, A_i is of the form $\forall x A_s$ for some individual variable x and some A_s, $s < i$ — or A_i results from some A_s and A_t, $s < i$, $t < i$, by *modus ponens* — that is, $A_s = A_t \supset A_i$ for some $s < i$ and $t < i$. If $\sigma = A_1, \ldots, A_i, \ldots, A_m$ is a deduction of P, then the Gödel number of σ will be $p_1^{g(A_1)} \cdot \ldots \cdot p_i^{g(A_i)} \cdot \ldots \cdot p_m^{g(A_m)}$. Now let a be any natural number other than 0 or 1, 0 and 1 not being the Gödel number of deductions in P. The prime factorization of a yields $a = p_1^{(a)_1} \cdot \ldots \cdot p_i^{(a)_i} \cdot \ldots \cdot p_{L(a)}^{(a)_{L(a)}}$. Hence, a will be the Gödel number of a deduction in the formal system P if and only if a does not equal 0 or 1 and, for all i less than or equal to the 'length' of a, L(a), the exponent of p_i in the prime factorization of a, $(a)_i$, is either the Gödel number of an instance of an axiom of P, the Gödel number of the wff that results from a wff whose Gödel number is among the $(a)_j$, $j < i$, by generalization, or the Gödel number of the wff that results from two wffs whose Gödel numbers are among the $(a)_j$, $j < i$, by *modus ponens*. The materially adequate demonstration that deduct is primitive recursive consists of the translation of the above analysis into symbols:

deduct(x) $\Leftrightarrow$ x $\neq$ 0 and x $\neq$ 1 and

$\forall i \leq L(x)$ {i = 0 or axiom$_1$((x)$_i$) or ... or axiom$_k$((x)$_i$) or

$\exists j < i \; \exists y < x$ (var(y) and (x)$_i$ = g($\forall$) * y * (x)$_j$)

or $\exists j < i \; \exists s < i$ ((x)$_s$ = g(( ) * (x)$_j$ * g($\supset$) * (x)$_i$ * g( )))}

where axiom$_1, \ldots,$ axiom$_k$ enumerate the axioms of P. As with the examples above, the primitive recursiveness of deduct is seen to depend on the prior demonstrations that the various functions and relations occurring on the right-hand side of the equation are primitive recursive and that the ways of constructing functions and relations used in the right-hand side of the equation preserve primitive recursiveness.

Working back from the construction of G in this manner illuminates the fact that the schedule of proofs provides for, and builds to, the eventual construction of G in the diagonalization/'proof'. However, in that this procedure of working backwards depends on the availability of the schedule of proofs as a completed object, it gives only an initial and retrospective sense of the 'directed' character of the lived-work of producing that schedule. I want to now indicate what the directness of that work actually consists of.

The discussion that follows will be facilitated by the availability of a referentially-specific definition of the syntax of a formal system. In

A SCHEDULE OF PROOFS

order to give such a definition, some preliminary terminology must be introduced. First, by a *syntactic definition*, I will refer to a definition identifying some feature of a formal system in which the definiendum, a *syntactic object*, is specified entirely in terms of ordered arrangements of the primitive symbols of that system. As is customary, the terms *'formation rules'* and *'axiomatics'*[18] will be used to refer to the hierarchies of syntactic definitions that are needed for — and that culminate, respectively, in — the definitions of the wffs and the *theorems* of that system. For the system P, the formation rules are given by specifying the concatenated sequences of primitive symbols that constitute the variables, the terms, and the wffs. The axiomatics of P include[19] the constellation of definitions surrounding the notions of substitution — namely, those concerning bound and free occurrences of individual variables, that of a term being free for a variable in a wff, and that of the operation of substitution itself — as well as the specification of the axiom schemata and rules of inference, the definition of a deduction, and the culminating definition, that of the *theorems* of P. Finally, by the syntax[20] of a formal system, I[21] will refer to the collection of syntactic objects that are defined by the formation rules and axiomatics of that system.

The reason for giving this definition of the syntax of a formal system is that it makes precise the notion that a Gödel numbering, temporarily considered as merely establishing a correspondence between the language[22] of a formal system and the natural numbers, renders the syntax of a formal system as a collection of numerical functions and relations. The syntax of a system so rendered will be referred to as the arithmetized syntax of that system, therein providing a descriptive definiteness to the vernacular expression that a Gödel numbering 'arithmetizes' the syntax of a formal system.

With these definitions in hand, let us return to the discussion of the 'directed' character of the schedule of proofs.

In the schedule of proofs outlined above, Propositions 1 through 12 represent a selection of the rudimentary facts concerning primitive recursive functions and relations that were well-known prior to Gödel's undecidability paper of 1931. Against the background of practices represented by these results, Propositions 13, 14, 15 and 16 are relatively easy to prove. However, the fact that they are relatively easy to prove offers no motivation internal to the work of proving Gödel's theorem for their introduction. That motivation comes from the fact that with the apparatus supplied by these propositions, the additional mathematical structure that the Gödel numbering gives to the arithmetized syntax of P is such that the prover can begin to write formulas seeably/showably defining the functions and relations of the arithmetized syntax as primitive recursive ones. More accurately, the need for the apparatus of Propositions 13 through 16 arises co-temporaneously

A SCHEDULE OF PROOFS

with the envisioned use of a numbering like a Gödel numbering as a technique of proving. With, for example, Gödel numbers a_1 and a_2 assigned to formal expressions A_1 and A_2, the attempt to write the formula for the Gödel number of $A_1 \supset A_2$ indicates the need for introducing an operation like that of *, thereby allowing the prover to write $g(A_1 \supset A_2) = a_1 * g(\supset) * a_2$.

From within the work of attempting to write, and of developing an apparatus for writing, the Gödel numbers of syntactic objects like that of $A_1 \supset A_2$ and, for example, like the individual variables, the increasing orderliness of that work exhibits the possibility that the entire syntax of P — rendered as numerical functions and relations under the Gödel numbering — can be shown to consist of primitive recursive functions and relations. Although, in the end, the project that is thus initiated cannot be completed, only the culminating relation, that of theorem(x),[23] cannot be shown to be primitive recursive. The part of the project that can be realized is sufficient for the construction of the diagonalization/'proof'.

The point is this: given the arithmetization of the syntax of P under a Gödel numbering, the presence and availability of the Gödel numbering as a technique of proving sets up the program of demonstrating that that arithmetized syntax consists of primitive recursive functions and relations. The notion of the 'directed' character of the schedule of proofs is a reference to this prospective enterprise as it unfolds in and as the work of producing the schedule of proofs.

3 *The selection and arrangement of these-particular propositions as composing this-particular, intrinsically sequentialized order of proving is the situated achievement of the work of producing the schedule of proofs*

The aim of the following four topics, (a)–(d), is to descriptively enrich and elaborate this proposal by recalling for the reader the lived-work of enunciating and organizing the schedule of proofs. In that the accountably-ordered schedule of proofs is the achievement of this lived-work, the fact that the finished schedule is an accountable order of proving — and can, therefore, as an example, be rendered as a structure of logical dependencies — does not explicate the schedule's temporally-situated and temporally-developing construction. Thus, by making the development of the schedule available, in its technical detail, as a production problem, the following material sets in relief the radical problem of specifying the coherence of the work of producing the schedule as that schedule is actually being constructed. This same material also provides a curious solution to that problem — that the coherence of the work (or of all the things that make up and are

A SCHEDULE OF PROOFS

spoken of as that 'coherence') inhabits and is the work itself.

(a) *Six themes concerning the lived-work of producing a schedule of proofs*
The major part of topic (a) will be devoted to the review of the construction of the 'concatenation' function * as the primitive recursive function,

$$x * y = x \cdot \prod_{i=0}^{L(y)} p_{L(x)+i}^{(y)_i}.$$

The idea will be to place that construction within the encompassing work of producing the schedule so as to find, in *'s construction, what it means to speak of that construction as being 'within' in the first place. In the course of adequately defining *, a significant portion of the schedule of proofs will be developed, and topic (a) will be brought to a close by extracting six summarizing themes from that material.

As a means of opening the discussion, let us begin by reviewing the proof that the set of Gödel numbers of terms of P is a primitive recursive relation. The reader will recall that I began by considering the particular term

$$t = +(x_1 S(\cdot(x_5 0))).$$

A formation sequence τ,

$$\tau = x_5, 0, \cdot(x_5 0), S(\cdot(x_5 0)), x_1, +(x_1 S(\cdot(x_5 0)))$$

was then associated with t. Corresponding, under the Gödel numbering, to this sequence of terms is a sequence of numbers, namely

$$g(x_5), g(0), g(\cdot(x_5 0)), g(S(\cdot(x_5 0))), g(x_1), g(+(x_1 S(\cdot(x_5 0)))),$$

from which the 'sequence number'

$$y = p_1^{g(x_5)} \cdot p_2^{g(0)} \cdot p_3^{g(\cdot(x_5 0))} \cdot p_4^{g(S(\cdot(x_5 0)))} \cdot p_5^{g(x_1)} \cdot p_6^{g(+(x_1 S(\cdot(x_5 0))))}$$

was finally constructed. The idea of this numbering was that it made available a methodic procedure, embedded in the techniques of constructing primitive recursive functions and relations, for determining whether or not a given number x was the Gödel number of a term, that procedure being articulated in the formula

$$\text{term}(x) \Leftrightarrow x \neq 0 \text{ and } \exists y \leq \prod_{i=0}^{L(x)} p_i^x \, \{[(y)_{L(y)} = x] \text{ and}$$

$$\forall i \leq L(y) \, ([i = 0] \text{ or } [(y)_i = g(0)] \text{ or }$$

$$[\text{var}((y)_i)] \text{ or } \exists j < i \, [(y)_i = g(S(\ )) * (y)_j * g(\ ))]$$

81

or $\exists j < i \; \exists k < i \; ([(y)_i = g(+(\;) * (y)_j * (y)_k * g(\;))]$ or
$[(y)_i = g(\cdot(\;) * (y)_j * (y)_k * g(\;))])\}$.

Now the prover, in beginning to construct a schedule of proofs, may not foresee the exact construction of such a formula nor may he anticipate the problematic details of such a construction for the particular, syntactically-specified theory in which he is working. On the other hand, in that the construction of such formulas is the projected thing to which his proving must come to, he can and does inspect the projected schedule as a way of informing the production of it. In this way, the prover may come to envision the following functions as comprising the apparatus for working with the Gödel numbering while, at the same time, recognizing that the adequacy of that apparatus will be determined over, and is contingent on, the course of working out the schedule itself:

(i) a set of 'encoding' functions, one for each natural number $m \geqslant 1$, such that

$\langle \; \rangle : N^m \to N$

$\langle \; \rangle : (a_1, \ldots, a_m) \mapsto \langle a_1, \ldots, a_m \rangle = p_1^{a_1} \cdot \ldots \cdot p_m^{a_m}$.

One of these functions maps, for example, the sequence of numbers

$g(x_5), g(\theta), g(\cdot(x_5\theta))), g(S(\cdot(x_5\theta))), g(x_1),$

$g(+(x_1 S(\cdot(x_5\theta))))$

corresponding to the formation sequence τ to the sequence number

$y = p_1^{g(x_5)} \cdot p_2^{g(\theta)} \cdot p_3^{g(\cdot (x_5 \theta))} \cdot p_4^{g(S(\cdot (x_5 \theta)))} \cdot p_5^{g(x_1)}$
$\cdot p_6^{g(+(x_1 S(\cdot (x_5 \theta))))}$

thereby 'encoding' the formation sequence τ in the number y.

(ii) a 'decoding' function $(\;)_\cdot$, mapping a given number x into the exponent of the serially-numbered $\cdot$-th prime in the prime factorization of x:

$(\;)_\cdot : N \times N \to N$

$(\;)_i : x = p_1^{a_1} \cdot \ldots \cdot p_i^{a_i} \cdot \ldots \cdot p_m^{a_m} \mapsto a_i.$

This function provides the device for inspecting each of the exponents in the prime factorization of a number to determine, for example, whether or not those exponents correspond to the Gödel number of terms comprising a formation sequence for t.

(iii) a 'concatenation' function. If A and B are sequences of primitive

A SCHEDULE OF PROOFS

symbols of P, then so is the expression AB formed by concatenating the symbols of A with those of B. Let $g(A) = p_1^{a_1} \cdot \ldots \cdot p_m^{a_m}$ and $g(B) = p_1^{b_1} \cdot \ldots \cdot p_n^{b_n}$. Then $g(AB) = p_1^{a_1} \cdot \ldots \cdot p_m^{a_m} \cdot p_{m+1}^{b_1} \cdot \ldots \cdot p_{m+n}^{b_n}$. The 'concatenation' function * maps $(g(A), g(B))$ to $g(AB)$, mirroring the process of concatenating the symbols, expressions, and sequences of expressions of P with an operation defined on pairs of numbers. Provisionally,[24] let us define * as

$* : N \times N \to N$

$* : (x, y) \mapsto x * y = p_1^{a_1} \cdot \ldots \cdot p_m^{a_m} \cdot p_{m+1}^{b_1} \cdot \ldots \cdot p_{m+n}^{b_n}$

where $x = p_1^{a_1} \cdot \ldots \cdot p_m^{a_m}$ and $y = p_1^{b_1} \cdot \ldots \cdot p_n^{b_n}$.

Then, as an example, the Gödel number of $A \supset B$ can be written $g(A \supset B) = g(A) * g(\supset) * g(B)$ once, of course, the associativity[25] of * has been established.

In addition to the projected construction of these functions, the prover also has, in beginning to construct the schedule of proofs, in and as the increasing articulation of the accountable work of that construction, that way of working as a familiar and remembered course-of-action. Thus, the prover may start the construction of the schedule by showing that first the divisibility relation, then the set of prime numbers, and then the function enumerating the primes are primitive recursive:[26]

( ) The 2-place numerical relation $x \mid y$ is primitive recursive.

Proof: $x|y \Leftrightarrow \exists n \leq y \, (y = n \cdot x)$.[27]

( ) Let prime(x) hold if and only if x is a prime number. Then prime(x) is a primitive recursive relation.

Proof: prime(x) $\Leftrightarrow 1 < x$ and

$\forall y \leq x \, (y|x \text{ implies } y = 1 \text{ or } y = x))$.[28]

( ) The function p_n giving, for each n, the n-th prime number is primitive recursive. ($p_0 := 1$.)

Proof: $p_0 = 1$

$p_{n+1} = \mu x \leq \square \, (\text{prime}(x) \text{ and } p_n < x)$.[29]

This is not to say that the writing of these propositions is automatic or mechanical. The prover, in a manner to be illustrated shortly, may have to find this or a similar order as an appropriate order for the propositions, or he may more or less remember the propositions and their order but, in the course of working out their proofs, come to establish and, therein, to see again that particular order as an accountable order of work. Even if the prover is using one or several previous proofs of Gödel's theorem as guides, he will still have to find for himself, over the course of writing their proofs, in the material details of that writing, that the selection and arrangement of the propositions given by those proofs is, in fact, an accountable order of proving. However, the aim of this discussion is neither to examine all the details of the schedule's construction — leaving aside the question of how the relevant and identifying details of the schedule emerge from within and are tied to the work of that construction; a question that I will return to in the next chapter — nor is it to assert that the construction of the schedule is done in some specific manner. The material that has just been outlined simulates the circumstances and material detail that might surround the prover as he undertakes the construction of the 'concatenation' function * as a primitive recursive function in the midst of constructing the schedule itself. It is with this material as background that I now want to turn to the examination of the work of constructing that function.

Let us begin that examination by reconsidering the definition of * that was given earlier:

$$x * y = p_1^{a_1} \cdot \ldots \cdot p_m^{a_m} \cdot p_{m+1}^{b_1} \cdot \ldots \cdot p_{m+n}^{b_n}$$

where $x = p_1^{a_1} \cdot \ldots \cdot p_m^{a_m}$ and $y = p_1^{b_1} \cdot \ldots \cdot p_n^{b_n}$. An immediately recognizable problem with this definition, independently of whether or not it defines a primitive recursive function, is that the numbers 0 and 1 do not have prime factorizations and, hence, that * is not well-defined. As a means of working toward this problem's solution, let us first consider the case when $x = 1$. By using the fact that 1 was defined as the 0-th prime p_0 — or, alternatively, by introducing such a modification into the definition of p_n at this point in the development of the schedule[30] — the prime factorization of any number $x = p_1^{a_1} \cdot \ldots \cdot p_m^{a_m}$ can be written as $x = p_0^s \cdot p_1^{a_1} \cdot \ldots \cdot p_m^{a_m}$ where s is an arbitrary natural number.[31] Both for definiteness and in order to see what it will come to in the work that follows, let us temporarily adopt the convention that $s = 0$. The aim of introducing this device of adding p_0^0 to the prime factorization of a number is that, by defining the prime factorization of 1 as p_0^0, the formula

$$x = p_0^0 \cdot p_1^{a_1} \cdot \ldots \cdot p_m^{a_m}$$

will hold, for some sequence of primes $p_0, \ldots, p_m$ and some sequence of exponents $0, a_1, \ldots, a_m$, for all $x > 0$, and, further, the formula for * can then be written as

$$x * y = p_0^0 \cdot p_1^{a_1} \cdot \ldots \cdot p_m^{a_m} \cdot p_r^0 \cdot p_{m+1}^{b_1} \cdot \ldots \cdot p_{m+n}^{b_n}$$

where r, in this case, is arbitrary.[32] For definiteness, let $r = 0$ although we will need to recall, at an appropriate time in our later work, that this choice was arbitrary and, therefore, that it can be changed. Provisional on the adequate definitional explication of our conventions in and as part of the developing schedule, we can define * as

$$x * y = p_0^0 \cdot p_1^{a_1} \cdot \ldots \cdot p_m^{a_m} \cdot p_0^0 \cdot p_{m+1}^{b_1} \cdot \ldots \cdot p_{m+n}^{b_n}$$

for all x and y greater than 0, $x = p_0^0 \cdot p_1^{a_1} \cdot \ldots \cdot p_m^{a_m}$, and $y = p_0^0 \cdot p_1^{b_1} \cdot \ldots \cdot p_n^{b_n}$, thus extending the initial definition of * to the cases when $x = 1$ and $y = 1$.

As a next step in developing the formula for *, let us consider the case when $x = 0$. In this case, x simply does not have a prime factorization — that is, $0 \neq p_0^0 \cdot p_1^{a_1} \cdot \ldots \cdot p_m^{a_m}$ for any possible sequence of prime numbers $p_0, \ldots, p_m$ and exponents $0, a_1, \ldots, a_m$. In the presence of this circumstance, we are forced to look for a different kind of modification of the formula for *. One possibility, which on its introduction appears as a 'natural' solution,[33] is to define * as

$$x * y = x \cdot p_0^0 \cdot p_{m+1}^{b_1} \cdot \ldots \cdot p_{m+n}^{b_n}$$

which yields $x * y = 0$ when $x = 0$ and which can be further clarified as

$$x * y = x \cdot p_0^0 \cdot \prod_{i=1}^{n} p_{m+1}^{b_i}.$$

That $x * y = 0$ when $x = 0$ can be interpreted as saying that the null, or empty, sequence of symbols of P concatenated with a non-null sequence B is a null sequence. Although, in this interpretation, the formula for * seems to lead to a false conclusion, the Gödel numbers of any member of Lg(P)[34] will always be greater than 0 or, for that matter, 1. Thus, in its projected use — as in writing $g(A \supset B) = g(A) * g(\supset) * g(B)$ where A and B are already identified as non-null sequences of symbols — the formula appears, at least for the moment, as being unproblematic.

Finally, let us address the case when $y = 0$. Here, even the last formula does not work. In view of the difficulties presented by this case, one alternative is to provisionally define * 'by cases,'[35]

$$x * y = \begin{cases} 0 \text{ if } y = 0 \\ x \cdot p_0^0 \cdot \prod_{i=1}^{n} p_{m+1}^{b_i} \text{ otherwise, where } x = 0 \text{ or} \\ \qquad x = p_0^0 \cdot p_1^{a_1} \cdot \ldots \cdot p_m^{a_m} \text{ and} \\ \qquad y = p_0^0 \cdot p_1^{b_1} \cdot \ldots \cdot p_n^{b_n} \end{cases}$$

and, thereby, to also further provide for the primitive recursiveness of *.

The sketch of the development of a formula for * that has been given to this point may strike the reader as the depiction of a fully 'rational' and 'orderly' course of inquiry and discovery that a prover of Gödel's

A SCHEDULE OF PROOFS

theorem might undertake and whose work would then come to fulfill. By this I mean that the reader may have understood the work of developing the last equation as being intrinsically connected to the sequential placement of the various approximating equations in the reasoned argumentation that I have used to motivate their introduction and modification. If this is so, a reinterpretation of the preceding discussion must be given. An actual prover will work his way to the last equation by *writing* formulas similar to the ones displayed and by *modifying* formulas already written — by crossing out symbols, adding symbols, inserting symbols with arrows, and the like. In doing so, he will provide for the recovery of the devices that are occasioned by his work (and that are being locally employed as prospective solutions of the problems that the written formulas manifest) in and as the found order and in-course rearrangement of his writings. In other words, the prover may articulate the course of reasoning that will lead again to the last equation 'only' in and as the accountable organization of the notes of his page of writings. To say this, however, is not to say that the prover is not always in the presence of the practically accountable, real thing that he is doing. Instead, it is intended to point to the local character of the 'rationality' of what the prover is doing as that 'rationality' inhabits the course of the prover's work and, therein, to point to the fact that *that* 'rationality' is betrayed by the devices of a 'descriptive narrative' like the one that I have given. Thus, for example, the prover will see in his introduction of p_0^0 in the expression $x * y = p_1^{a_1} \cdot {}_0 \cdot \cdot {}_b p_m^{a_m} \cdot p_{m+1}^{b_1} \cdot \ldots \cdot p_{m+n}^{b_n}$ (i.e., $x * y = p_0^0 \cdot p_1^{a_1} \cdot \ldots \cdot p_m^{a_m} \cdot p_0^0 \cdot p_{m+1}^{b_1} \cdot \ldots \cdot p_{m+n}^{b_n}$) the purposeful thing that he is doing, and he will also find that device as a reflectively uninteresting, natural consequence of those doing — as something not needing to be specifically elaborated as an appropriate, adequate, or efficacious procedure disengaged from the further material development of the definition of *. Similarly, the occasioned need for introducing p_0^0 need not develop in the fashion that I have narrated, but could be motivated by considerations of the manipulatable things that can be done with formulas already on the page of working notes.

These matters will be addressed again later. For the moment, let us return to the construction of *.

In that the prover of Gödel's theorem knows[36] that the problematic character[37] of the initial definition of * when x or y equals 0 or 1 can be circumvented by defining * by cases, the prover will also anticipate that the problems arising from these cases are not serious for the function *-proper that is intended to mirror the concatenation operation on Lg(P) with an operator on N X N. In fact, immediately following his writing the preliminary definition and seeing that definition's inadequacies, the prover may, as part of a natural course of reasoning, come to inspect the projected use of * to discover whether or not

A SCHEDULE OF PROOFS

those cases are consequential to its eventual, adequate definition.
  A potentially more serious problem in defining * comes from the fact that the prover must be able to determine[38] a natural number's prime factorization as a primitive recursive function of that number in order to modify the formula

$$x * y = p_1^{a_1} \cdot \ldots \cdot p_m^{a_m} \cdot p_{m+1}^{b_1} \cdot \ldots \cdot p_{m+n}^{b_n}$$

or

$$x * y = x \cdot p_0^0 \cdot \prod_{i=1}^{n} p_{m+i}^{b_i}$$

so as to seeably/showably define * as itself a primitive recursive function. In particular, one of the functions that will be needed is one that gives the 'length' of a natural number or, more precisely, that gives the serial-number[39] of the largest prime in a given number's prime factorization. Let this function be denoted by L. Then, if $x = p_1^{a_1} \cdot \ldots \cdot p_m^{a_m}$ and $y = p_1^{b_1} \cdot \ldots \cdot p_n^{b_n}$, the prover wants $L(x) = m$ and $L(y) = n$. With this function in hand, he will be able to write

$$x * y = x \cdot p_0^0 \cdot \prod_{i=1}^{L(y)} p_{L(x)+i}^{b_i},$$

bringing the definition of * closer to exhibiting its primitive recursiveness.
  Once again, there is something slightly deceptive in the way of speaking of the discovered need for the function L. It is on the occasion of the need for such a function — as, for example, when the prover is surveying a preliminary definition of * like

$$x * y = p_1^{a_1} \cdot \ldots \cdot p_m^{a_m} \cdot p_{m+1}^{b_1} \cdot \ldots \cdot p_{m+n}^{b_n}$$

to see what can be appropriately modified about that definition — that the prover will recall the presence of just such a function/device in the proofs of Gödel's theorem with which he is familiar.[40] However, after seeing the materially purposeful thing that the introduction of L could do, the prover may not set about constructing a formula defining it. Instead, the prover may defer that construction and first use L to develop the formula for * as

$$x * y = x \cdot p_0^0 \cdot \prod_{i=1}^{L(y)} p_{L(x)+i}^{b_i}$$

and, using the 'decoding' function $(\ )_.$, further or co-temporaneously[41] develop it even as

$$x * y = x \cdot \prod_{i=0}^{L(y)} p_{L(x)+i}^{(y)_i},$$

therein producing a seeably/showably primitive recursive definition of * subject to the adequate definition of L and $(\ )_.$. If, in fact, he works in this manner, he will also note, over the course of the work of writing such formulas, the proper reorganization of that work as the accountable work of the locally-obtained, projectively final version of that part of the schedule of proofs on which he is currently working.

What is needed to define L as a seeably/showably primitive recursive function? Well, what is $L(x)$? We want $L(x)$ to be the number of the largest prime that divides x or, in terms compatible with the ways of constructing primitive recursive functions that are presently available, we want it to be the number m such that $p_m | x$ but $p_k \nmid x$ for all $k > m$. As a means of obtaining that number m, we can use the least number operator:

$$L(x) = \mu m \leqslant \Box \, (p_m | x \text{ and } \forall k \leqslant \Box \, (m < k \text{ implies } p_k \nmid x)).^{42}$$

By supplying proper upper bounds for m and k, this formula becomes

$$L(x) = \mu m \leqslant x \, (p_m | x \text{ and } \forall k \leqslant x \, (m < k \text{ implies } p_k \nmid x)),$$

defining a primitive recursive function, the 'length' of x. Furthermore, by checking this definition for the potentially problematic cases $x = 0$ and $y = 0$, one finds that $L(0) = 0$ and $L(1) = 0$ as they 'should'.[43]

In that L is needed for later propositions of the schedule (as, for example, in the formula defining $term(x)$) and in that the serial character of the schedule is to be maintained (as part of what will later come to be called the schedule's 'structure of proving'), the proposition that L is primitive recursive is enunciated as a separate proposition of that schedule. In that that proposition has been found to be needed for the proof that $x * y$ is primitive recursive, it is also found to properly precede the proposition concerning $x * y$ in the 'finished' schedule of proofs.

'Next,' let us consider an appropriate modification of our new formula

$$x * y = x \cdot p_0^0 \cdot \prod_{i=1}^{L(y)} p_{L(x)+i}^{b_i}$$

so as to obtain, in a primitive recursive manner, the exponents $b_1, \ldots, b_n$ of the prime factorization of y. By using the 'decoding' function $(\;)_{\bullet}$ with $b_1 = (y)_1, b_2 = (y)_2, \ldots, b_n = (y)_{L(y)}$, this formula can be rewritten as

$$x * y = x \cdot p_0^0 \cdot \prod_{i=0}^{L(y)} p_{L(x)+i}^{(y)_i}$$

and the need for $(\;)_{\bullet}$ will, therein, either be recalled or, if its use was anticipated, be recalled again in and as the first instance demanding its articulation and use.

A definition of $(x)_i$ as a primitive recursive function of x and i can be found as

A SCHEDULE OF PROOFS

$(x)_i = \mu k \leq x \,(p_i^k \mid x \text{ and } p_i^{k+1} \nmid x).$[44]

Checking this formula for the potentially problematic cases $x = 0$, $x = 1$, and $i = 0$, one finds that $(1)_i = 0$ for all $i > 0$, but for the cases when $i = 0$ or when $x = 0$, the formula is 'apparently' not well-defined.[45] The solution of this problem may come about as the 'discovered gestalt'[46] of the following circumstances: By reviewing the work of coming to the formula

$$x * y = x \cdot p_0^0 \cdot \prod_{i=1}^{L(y)} p_{L(x)+i}^{(y)_i}$$

as a feature of inspecting that formula to find a next thing to be done, the prover will find again (as something already known) the arbitrariness of the selection of p_0^0 in that formula. Seeing, 'therefore,' that he can write $p_{L(x)}$ in the place of p_0, the prover can obtain

$$x * y = x \cdot p_{L(x)}^0 \cdot \prod_{i=1}^{L(y)} p_{L(x)+1}^{(y)_i}$$

which is appropriate, by definition of $L(x)$, even when $x = 0$ and $x = 1$. If $(y)_0$ is defined as equaling 0 for all y, then $p_{L(x)}^0 = p_{L(x)}^{(y)_0}$ and

$$x * y = x \cdot p_0^0 \cdot \prod_{i=1}^{L(y)} p_{L(x)+i}^{(y)_i}$$

$$= x \cdot p_{L(x)}^{(y)_0} \cdot \prod_{i=1}^{L(y)} p_{L(x)+i}^{(y)_i}$$

$$= x \cdot \prod_{i=0}^{L(y)} p_{L(x)+i}^{(y)_i},$$

the last line defining * for all $x, y \in N$ with $x * y = 0$ if $x = 0$ and with $x * y = x$ if $y = 0$. This being the case, the prover will define $(x)_0 := 0$ for all x, implying $(0)_0 := 0$ as well. The appropriate definition of $(\ )_.$ therein emerges as

$$(x)_i = \begin{cases} 0 \text{ if } x = 0 \text{ or } i = 0 \\ \mu k \leq x \,(p_i^k \mid x \text{ and } p_i^{k+i} \nmid x) \text{ otherwise} \end{cases}$$

and with it, *'s definition as well:

$$x * y = x \cdot \prod_{i=0}^{L(y)} p_{L(x)+i}^{(y)_i},$$

a seeably/showably primitive recursive function of x and y. In particular, $x * y$ is defined when $y = 0$, and the formula is applicable without introducing special conventions for the prime factorization of 1. Furthermore, given this way of working, the proofs of the primitive recursiveness of ( ) and L naturally precede that of *, and the proofs of ( ) and L are found to be independent of one another and, therefore, arbitrarily ordered in relation to each other. In that the definition of * depends on that of L, the definition of L 'should' directly precede that of *.

One last problem with the formula for * remains, although that problem may make its appearance only later in the development of the schedule. The problem is this: For almost all of the proofs of the schedule concerning the arithmetized syntax of P, a prover (heuristically)[47] uses the *-operator to exhibit the 'structure' of a natural number x by envisioning that number as the Gödel number of a member of a particular subset of Lg(P) and then 'decomposing' it into its constituent parts. For example, a prover will want to write formulas like x = g(A) * g($\supset$) * g(B) exhibiting the fact that if g(A) and g(B) are the Gödel numbers of wffs A and B of P, then x is the Gödel number of the wff $A \supset B$. Now the ordered sequence of primitive symbols of P, $\alpha_1 \ldots \alpha_p$, that is being meta-syntactically abbreviated as $A \supset B$ — as, for example, if $A = \alpha_1 \ldots \alpha_{k-1}$, $\supset = \alpha_k$, and $B = \alpha_{k+1} \ldots \alpha_p$ — remains the same whether that ordered sequence is considered as the sequence of symbols making up $A \supset$ concatenated with the ordered sequence of symbols identified as B, or it is considered as the symbols of A concatenated with those of $\supset B$, or the sequence of primitive symbols is considered as being partitioned in some other way. The prover wants a similar property to hold for the *-operator; for example, he will want to be able to unambiguously write

x = g($A \supset B$) = g(A) * g($\supset$) * g(B)

independently of whether g($A \supset B$) is decomposed as

g($A \supset B$) = {g(A) * g($\supset$)} * g(B)

or as

g($A \supset B$) = g(A) * {g($\supset$) * g(B)} .

Formulated more generally, the property of * that needs to be established is its associativity — that is, that

(x * y) * z = x * (y * z)

for all x, y, z $\in$ N.

In a sense, the associative property of * *on the set of Gödel numbers*[48] is already ensured by the fact that the definition of * articulates the methodic character of the Gödel numbering and the Gödel numbering was constructed so as to assign unique numbers to different elements of Lg(P) independently of the various ways in which those elements can be constructed by concatenating the primitive symbols that compose them. The prover, however, on the occasion of questioning the associativity of *, will not necessarily reflect on the reasonableness of * being associative, but, instead, will 'simply' examine[49] *'s definition

$$x * y = x \cdot \prod_{i=1}^{L(y)} p_{L(x)+1}^{(y)_i}$$

to see if, in fact, that property holds. Such an examination locates the one problematic case as being $y = 0$. Using *'s definition, the prover will come to calculate

$(x * 0) * z = x * z$

$x * (0 * z) = z.$

In that the equality

$x * z = x$

does not generally hold, the prover will, therefore, not be able to assert, without qualification, that * is associative.

Two solutions to this problem seem to be available. As I suggested earlier, the prover could redefine * by cases, as in

$$x * y = \begin{cases} 0 \text{ if } y = 0 \\ x \cdot \prod_{i=0}^{L(y)} p_{L(x)+i}^{(y)_i} \text{ otherwise} \end{cases}$$

thereby ensuring the associativity of * by defining $x * y$ to always equal 0 when either x or y equals 0. Such a device is somewhat artificial: the formula $x \cdot \prod_{i=0}^{L(y)} p_{L(x)+i}^{(y)_i}$ already adequately defines * as a primitive recursive function on $N \times N$, and, in that this is made available to the reader by offering $y = 0$ as a separate case, the distinction between $y = 0$ and $y \neq 0$ in a definition of * by cases instigates an examination and discussion of how * will actually be used in the proofs of the schedule that follow its introduction. However, in contrast to this artificiality but tied to the examination of which it speaks, the definition of * by cases has a potentially serious consequence for the development of the schedule: in that the values of * are being chosen for some of its arguments the prover must look to the projected use of * to see if such a choice will affect the proofs that follow it.

A second solution to the problem raised by the desired associativity of * is afforded by the fact that, since all Gödel numbers are greater than 0 and since * will be applied only to Gödel numbers, the associativity of * need only be established when x, y, and z are all greater than 0. The prover — for example, in examining the justification for modifying the definition $x * y = x \cdot \prod_{i=0}^{L(y)} p_{L(x)+i}^{(y)_i}$ when $y = 0$, at the

A SCHEDULE OF PROOFS

time of that examination — may come to see, as a pointed relevance for his immediate work, that all Gödel numbers are greater than 0 and, therein, that the case when y = 0 is actually of no consequence for this later work. Rather than modifying the definition of * and motivating that modification for the reader, the prover will simply define * by the formula

$$x * y = x \cdot \prod_{i=0}^{L(y)} p_{L(x)+i}^{(y)_i}$$

and then note, in one way or another, that while the associativity of * will be needed later in the schedule, that associativity need only be established when x, y, and z are all non-zero.[50,51]

Let us now — i.e., at this point in the construction of the schedule of proofs — look back at the apparatus that was originally envisioned as being needed to show that the arithmetized syntax of P is made up of primitive recursive functions and relations. Of that apparatus, only the 'encoding' functions

$$\langle \rangle : (a_1, \ldots, a_m) \mapsto \langle a_1, \ldots, a_m \rangle = p_1^{a_1} \cdot \ldots \cdot p_m^{a_m}$$

have yet to be shown to be, or reconstructed as, or shown not to be primitive recursive. But here, the displayed equation already seeably/showably defines a primitive recursive function: the formula for $\langle a_1, \ldots, a_m \rangle$ can be rewritten as[52]

$$\langle a_1, \ldots, a_m \rangle = p_1^{a_1} \cdot \ldots \cdot p_m^{a_m}$$
$$= p_{Z_1^m(a_1,\ldots,a_m)}^{I_1^m(a_1,\ldots,a_m)} \cdot \ldots \cdot p_{Z_i^m(a_1,\ldots,a_m)}^{I_i^m(a_1,\ldots,a_m)} \cdot \ldots \cdot p_{Z_m^m(a_1,\ldots,a_m)}^{I_m^m(a_1,\ldots,a_m)}$$

where I_i^m is, as the reader will recall, the projection function mapping $(a_1, \ldots, a_i, \ldots, a_m)$ onto its i-th coordinate a_i and Z_i^m is a situationally-occasioned notational 'innovation'[53] for the function that extends the domain of the primitive recursive function $Z_i(a) = i$ from N to N^m; in that the relevantly exhibited and exhibitable 'component' functions I_i^m, Z_i^m, exponentiation, the mappings

$$(a_1, \ldots, a_m) \mapsto Z_i^m(a_1, \ldots, a_m) \mapsto p_{Z_i^m(a_1,\ldots,a_m)} = p_i$$

$i = 1, \ldots, m$, and the finite product

$$(a_1, \ldots, a_m) \to a_1 \cdot \ldots \cdot a_m$$

are primitive recursive,[54] then so is $\langle a_1, \ldots, a_m \rangle$.[55] The immediate point is that the prover, in seeing how to go about showing that $\langle a_1, \ldots, a_m \rangle = p_1^{a_1} \cdot \ldots \cdot p_m^{a_m}$ defines a primitive recursive function, also sees what is (evidently) needed for such a demonstration and, therein, as a pointed relevance of that inspection, finds that the proof of the primitive recursiveness of $\langle \ \rangle$ is not dependent on the

A SCHEDULE OF PROOFS

primitive recursiveness of L, ( )., or *; that the proposition asserting the primitive recursiveness of ⟨ ⟩ may be placed irrespectively of the relative positioning of the assertions concerning L, ( )., or *; and, in that L, ( )., and * do exhibit dependencies among their proofs[56] and, therein, in that the produced arrangement of those propositions and their proofs compose an orderly course of proving among themselves, that the proposition concerning ⟨ ⟩ 'should not' be inserted within the arrangement of L, ( )., and *.

Let me give the schedule of proofs as it has been developed to this point:[57]

( ) The 2-place relation $x|y$ is primitive recursive.

Proof: $x|y \Leftrightarrow \exists n \leq y \, (y = n \cdot x)$

( ) Let prime(x) hold if and only if x is a prime number. Then prime(x) is a primitive recursive relation.

Proof: $\text{prime}(x) \Leftrightarrow x > 1$ and

$$\forall y \leq x \, (y|x \text{ implies } (y = 1 \text{ or } y = x)).$$

( ) The function p_n giving, for each n, the n-th prime number is primitive recursive. ($p_0 := 1$.)

Proof: $p_0 = 1$

$$p_{n+1} = \mu x \leq (p_n)^n + 1 \, \{\text{prime}(x) \text{ and } p_n < x\}.$$

To see that $(p_n)^n + 1$ is an upper bound on x, it is enough to note that $(p_1 \cdot \ldots \cdot p_n) + 1$ either is a prime number or is divisible by some prime number greater than p_n, for it then follows that $p_{n+1} \leq (p_1 \cdot \ldots \cdot p_n) + 1 < (p_n)^n + 1$.

( ) Define $(x)_n$ as the exponent of p_n in the prime factorization of x if $x > 1$ and $n > 0$ and as 0 otherwise.* Then $(x)_n$ is a primitive recursive function of x and n.

Proof:
$$(x)_n = \begin{cases} 0 \text{ if } x = 0 \text{ or } i = 0 \\ \mu k \leq x \, (p_n^k | x \text{ and } p_n^{k+1} \!\!\not|\, x) \text{ otherwise} \end{cases}$$

( ) Let the function L(x) give the number n of the largest prime p_n in the prime factorization of x or give 0 if x is 0 or 1.** L(x) is a primitive recursive function.

Proof: $L(x) = \mu n \leq x \, (p_n | x \text{ and } \forall k \leq x \, (n < k \text{ implies } p_k \!\!\not|\, x))$

( ) For every natural number y, $y = 0$ or $y = p_0^{(y)_0} \cdot p_1^{(y)_1} \cdot \ldots \cdot p_{L(y)}^{(y)_{L(y)}}$.

---

* E.g., $(294)_4 = (2 \cdot 3 \cdot 7^2)_4 = (p_1^1 \cdot p_2^1 \cdot p_3^0 \cdot p_4^2)_4 = 2$.

** E.g., $L(294) = L(p_1^1 \cdot p_2^1 \cdot p_3^0 \cdot p_4^2) = 4$. L(x) defines the 'length' of x.

A SCHEDULE OF PROOFS

Define x * y as the function mapping (x, y) to the value

$$x * y = x \cdot \prod_{i=1}^{L(y)} p_{L(x)+i}^{(y)_i} .†$$

Then x * y is a primitive recursive function of x and y.

It follows from the definition that x * (y * z) = (x * y) * z for all x, y, and z greater than 0. Thus, the finite 'product' of numbers $a_i > 0, i = 1, \ldots, n$, can be written unambiguously as $a_1 * \ldots * a_n$.

[58]( ) For each $n > 0$, define $\langle x_1, \ldots, x_n \rangle$ by the equation

$$\langle x_1, \ldots, x_n \rangle = p_1^{x_1} \cdot \ldots \cdot p_n^{x_n}.‡$$

Then $\langle x_1, \ldots, x_n \rangle$ is a primitive recursive function of $(x_1, \ldots, x_n)$.

I now want to bring topic (a) to a close by developing six themes that summarize and enrich the preceding discussion.

(i) There are two related ways in which the part of the schedule that was presented above need not have been constructed in the manner I have described. First, both the preceding text and the accompanying footnotes have already emphasized the variations of materially-specific and materially-motivated reasoning that could lead to the same, or to a recognizably similar, schedule of proofs. Second, a naturally accountable schedule of proofs need not, in the end, be identical to the schedule that was partially constructed. Thus, were a prover to have envisioned the need for[59] the primitive recursive functions[60]

$$sg(x) = \begin{cases} 1 \text{ if } x = 0 \\ 0 \text{ if } x \neq 0 \end{cases}$$

$$\overline{sg}(x) = \begin{cases} 0 \text{ if } x = 0 \\ 1 \text{ if } x \neq 0 \end{cases}$$

$$|x - y| = (x \dotminus y) \dotminus (y \dotminus x)$$

and had he 'similarly'[61] introduced/exhibited the function rm(x, y) giving the remainder upon division of y by x as a primitive recursive function,

---

† E.g., $294 * 6 = (2 \cdot 3 \cdot 7^2) * (2 \cdot 3) = (p_1^1 \cdot p_2^1 \cdot p_3^0 \cdot p_4^2) * (p_1^1 \cdot p_2^1)$
$= p_1^1 \cdot p_2^1 \cdot p_3^0 \cdot p_4^2 \cdot p_5^1 \cdot p_6^1$
$= 2 \cdot 3 \cdot 7^2 \cdot 11 \cdot 13 = 42{,}042.$

‡ E.g., $\langle 1, 1, 0, 2 \rangle = p_1^1 \cdot p_2^1 \cdot p_3^0 \cdot p_4^2 = 2 \cdot 3 \cdot 7^2 = 294.$

A SCHEDULE OF PROOFS

$rm(x, 0) = 0$

$rm(x, S(y)) = [S(rm(x, y))] \cdot [\overline{sg}(|x - S(rm(x, y))|)]$,

'then,'[62] writing K_R for the characteristic function of a numerical relation R, the proofs of the primitive recursiveness of $x|y$ and prime(x) could have been given as[63]

( ) The 2-place numerical relation $x|y$ is primitive recursive.

Proof: $K_|(x, y) = sg(rm(x, y))$

( ) Let $D(x)$ be the number of divisors of x if $x > 0$ and be 1 if $x = 0$. Then $D(x)$ is a primitive recursive function.

Proof: $D(x) = \sum_{i=0}^{x} sg(rm(i, x))$

( ) Let prime(x) hold if and only if x is a prime number. Then prime(x) is a primitive recursive relation.

Proof: $K_{prime}(x) = sg[(D(x) \dotdiv 2) + \overline{sg}(|x - 1|) + \overline{sg}(x \dotdiv 0)]$

It is clear, then, that the materially-specific detail and the materially-exhibited reasoning of any one such schedule may differ from other practically objective, comparably adequate schedules of proofs. But the ethnography that I gave earlier did not attempt to document either invariant material specificities of all schedules of proofs or invariant 'cognitive processes,' nor did that ethnography have such invariant descriptions as its projected goal. Instead, that ethnography — and the preceding example as well — make available, as an inspectable and researchable phenomenon, the finding that the work of producing a schedule of proofs is, in every particular case, from within that work itself, constrained by the material character of its own developing argument and that it is always from within just-this materially specific, just-this increasingly articulated, endogenously and developmentally organized way-of-working/mathematical object that a prover will further articulate and organize, as the endogenous work of his proving, both that object and, simultaneously, that way of working.

(ii) Once again, let me begin by making several observations. First, even though the lived-work of producing a schedule of proofs is, in each particular case, constrained by the material character of *its* developing argument, there is nothing about a particular way of working, from within the developing course of the work itself, that is self-exhibiting of its uniqueness in providing a solution for the problem-at-hand. As a feature of proving's work, the proof of a theorem can, potentially, always be given in a different and hitherto unrealized manner. Second, the solution of an unsolved problem — as the very thing that such a projected solution has come to be in that the problem is recognizably unsolved and, therein, not yet adequately understood — cannot be specified before it is found, nor can the existence of a

A SCHEDULE OF PROOFS

solution be guaranteed to be made accessible through any particular way of working. And third, as a prover develops a 'known' proof of a theorem (as, for example, on the request of a student), the familiar efficacy of 'the way of the theorem's proof' can be betrayed by the 'adversity' of the self-same, notationally-specific, temporally developing, endogenously organized manner of proving — that 'adversity' consisting, for example, of the further work that the prover's current methods recognizably project or of the curious 'loss' of the orderliness of the argument and, therein, simultaneously, the 'loss' of the vision of the thing that needs to be proved. In the face of such circumstances, the prover will search for alternative ways of proving that will circumvent (what are available as particularly-his) present troubles.

These observations lead to and set in relief a further observation, one made available by the ethnography as well: when a mathematician is working on a problem, he is actually searching for and cultivating 'something,' in and as the developing and projected writings on the working-page or blackboard, which, when found, makes up the thing that can then be spoken of as a mathematical discovery. Speaking particularly of the work of producing a schedule of proofs, what I propose is this: the lived-work of producing a schedule of proofs constructs that schedule, as its accomplishment, as an accountable course of mathematical proving, and it is that accountable course of proving, as it has come to be embodied in the schedule over the course of the schedule's construction, that is the thing that is being looked for as, and that then makes up, the mathematical discovery[64] (of that schedule). Furthermore, and in consequence, in that that discovery consists of the just-this endogenous organization of work practices, *that* discovery is irremediably and exclusively available in and as a local enterprise.

(iii) Consider now the 'finished' portion of the schedule of proofs that was given above. In the earlier discussion of the 'directed' character of a schedule of proofs, I began that discussion by indicating how the ordering of the propositions of a schedule could be graphically represented as 'lines of dependence,' either as

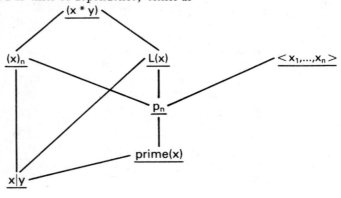

A SCHEDULE OF PROOFS

or, replacing the functions and relations with the associated numbers of the propositions of the schedule concerning them, as

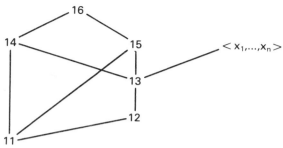

On being presented with such a graph, a prover will find in its depiction the availability of a method of representing the orderliness of the entire schedule of proofs,[65] and this, partly because that method does not seem to be peculiar to the schedule itself. The graph, in its disclosure as a method or representation, provides for essential and inessential features of the-work-of-the-schedule's-production/the-schedule's organization, and, therein, provides as well for a transcendental or platonic schedule of proofs that is disengaged from the local work of its production. The graph, together with its associated method, renders[66] the schedule of proofs as what could be called a 'structure of logical dependencies.'

In miniature, let us suppose that A, B and C are propositions of the schedule and that proposition A is used to prove proposition B which, in turn, is used to prove proposition C. Then the ordering of these propositions could be represented as A → B → C, where the arrows provide a similar function as the serially ordered numbers and 'reasonably' oriented line segments in the graph above. The immediate point is that to make the observations that A, B and C are needed in the schedule; that A, B and C adequately articulate a portion of the schedule and that the three make up a practically distinguishable and practically separable sequence of proofs; that A is used to prove B, that B is used to prove C, and that A is not directly used to prove C; that B is a necessary intermediate proposition between A and C; that the propositions are appropriately arranged as first A, then B, and then C: that A is, in fact, actually needed to prove B, and that B is actually needed to prove C — to make these observations and, thereby, to come to find the adequacy and cogency of A → B → C as a representation of the orderliness of a part of the schedule — already places the prover within the 'perspective of the proof's work.' In this way, then, the 'analysis' of the orderliness of the propositions of the schedule that is provided by A → B → C is already tied to the availability of the lived-work of proving just-those propositions; in that the propositions of the schedule are already

A SCHEDULE OF PROOFS

chained to their associated proofs and in that together, those propositions and proofs make up a naturally accountable course of proving, the enunciation and arrangement of the propositions thereby comes to exhibit, via this rendering, a transcendental order of work. The prover does not find the orderliness of the 'completed' schedule by first, if ever, consulting the logical requirements for such an order, but, instead, finds that orderliness by returning those propositions to the work of their proofs from within which they then take on their order properties. Thus, to draw 'lines of dependence', a prover must first find the associated work of the schedule's production; to see that proposition B is needed to prove proposition C is already to find oneself within the work of proving that the schedule both provides and is a part.

The larger point is this: in that a schedule of proofs *is* the endogenously organized, temporally and materially developing construction that it has come to be over the course of its production, that schedule takes on — not just its order properties as they become available through a schematic rendering of its 'logical structure' — but *all* of its properties as the naturally accountable object that it demonstrably is.

(iv) Today, if a mathematician were to go about proving-again Gödel's theorem, and were he to do so by explicitly constructing a schedule of proofs, it is unlikely that he would actually go through the detailed work of constructing such a schedule that I indicated and outlined above. Instead, he would probably construct a schedule of proofs in immediate consultation with established texts giving such schedules. This being the case, what then is the relevance of the preceding ethnography as a description of mathematicians' work? The relevance is this: although a mathematician might not undertake the construction of a schedule of proofs without consulting, over the course of his work, other, already 'completed' schedules, and although he might not, thereby, work out for himself the material detail and organization of the schedule, a mathematician is completely familiar, in and as his own work as a practising mathematician, of what 'finding' and, therein, articulating and organizing, mathematical proofs consists of, in its material detail, *as lived-work*. In viewing another prover's schedule of proofs or in juxtaposing a number of such schedules so as to find what needs to be done for the particular schedule that he himself is building, the mathematician reads through those schedules to find — and in reading them, finds — what they make available as the accountable work of producing just-their schedules of proofs, respectively. A mathematician need not go through the entire reconstruction of a given schedule to have access to that schedule as an adequate summary of a naturally accountable way of proving, and this is so not because of what the mathematician 'knows,' but because of what he is utterly familiar with as the endogenously organized workings of his own praxis.

A SCHEDULE OF PROOFS

The ethnography that I gave earlier descriptively explicates this familiar, yet, for the mathematician, unremarkable work in the particular case of the construction of a part of a schedule of proofs.

(v) Point (iv) lends itself to the discussion of another and, as the reader will see, related issue. In order to introduce it, let me begin by offering two different ways of speaking abut the 'received' history of a mathematical theorem.[67] The first and a 'weak' use of this notion is that a theorem and its proof are available to a mathematician (working in the particular field in which the theorem occurs) not just as a theorem and its proof, but as a theorem and proof that arise from within and that are responsive to the proper history of their discovery and development. That history is understood to include the accountable historical development of the methods that are used in the proof, it provides the accountable context from within which the theorem took, and now takes on, its importance, it provides an accountable history for the development of the current and alternative statements of the theorem, of its current and alternative proofs, and of its current uses in the field, it is available as the history of the work of particular people, etc.

The second and 'strong' use of the notion of the 'received' history of a theorem refers, in contrast to the preceding, to the 'state of the art' of the theorem's proof. Under this usage, the 'received' history of a theorem is not seen to consist of a proper history, but of 'received' ways of proving. The reason for this usage being 'strong' and the other usage being 'weak' is that it is by reason of the availability to a mathematician of that 'state of the art' that allows him to examine and construct an appropriate history of the theorem in question. It is in the presence of naturally accountable proofs of the theorem that a mathematician is able to find and argue what the proper history of that theorem and its proof is.

Having distinguished these two different ways of speaking about the 'received' history of a theorem, let us return to the discussion of the work of producing a schedule of proofs. The connection between the two is this: in that, over the course of constructing a schedule of proofs, a prover consults other 'completed' schedules as a means of constructing his own, a prover finds as an incidental feature of that consultation, what the 'received' history of the proof of Gödel's theorem — in the strong sense — at least in part, consists of. And, at the same time, a prover finds the place of his own work within that 'received' history as well. The point, then, is this: it is in this way — in that a schedule of proofs, over the course of its production, is constructed in consultation with already 'completed' schedules of proofs — that the material presentation and organization of such schedules are preserved, repeated and modified as appropriate, efficacious and accountable ways of proving Gödel's theorem.

(vi) Finally, it is now possible to begin to make descriptively precise what it means to speak of the *intrinsic* orderliness and the *intrinsic* sequentialization of the schedule of proofs. The idea is this: on *every* occasion when a prover comes to examine a schedule of proofs to see if the definitions of that schedule 'fit' together or to review the arrangement of the propositions or, in general, to find the accountable orderliness of a given schedule of proofs, the prover does so by returning those 'questions' to the work of producing the schedule that the given schedule itself makes available. And, in fact, in the very way that they arise, those 'questions' are already tied to, occur from within, and are not disengageable from the work of producing the schedule of proofs. To see, for example, that L(0) *should* equal 0 or to see that certain propositions *should* be arranged in such and such an order is not to find, by reflection on the properties of natural numbers, that 0 has no prime factorization and, 'therefore,' that $L(0) = 0$, nor is it to find that the nature of logical inference 'requires' a specific organization of propositions. Instead, to 'understand' the appropriateness of the definitions and the organization of a schedule is to find oneself, as a prover, inextricably engaged in the work of proving Gödel's theorem. To speak of the intrinsic orderliness of a schedule of proofs is to refer to the fact that the orderliness of a schedule of proofs is exhibited in and as a course of mathematical proving.

(b) *The construction of a schedule of proofs so as to provide an apparatus within itself for the analysis of the work of its own construction*

As part of the work of 'working through' an established schedule of proofs, a prover will use and interrogate the previous propositions of that schedule so as to provide an exegesis (and, therein, to establish the adequacy) of a given proof as a definition of a primitive recursive function or relation. Consider, for example, the following proof of the primitive recursiveness of the divisibility relation (Proposition 11):

$x|y \Leftrightarrow \exists n \leq y \, (y = n \cdot x)$.

On 'coming to' this formula while working through another prover's schedule, a prover might[68] thereupon compose — as a temporally realized, materially exhibited course of reasoning — a sketch derivation similar to the one below:

$n \cdot x$ (Proposition 3)

$y = z$ (Proposition 5)

$y = n \cdot x$ (Proposition 7)

$\exists n \leq w \, (y = n \cdot x)$ (Proposition 8)

$\exists n \leq y \, (y = n \cdot x)$ (Proposition 2)

the reasoning of which can be elaborated as follows: in that multiplication, n · x, and the equality relation, y = z, are, respectively, a primitive recursive function (by Proposition 3) and a primitive recursive relation (by Proposition 5), the equation y = n · x defines a primitive recursive relation in that it results from the 'substitution'[69] of a primitive recursive function for a variable in a primitive recursive relation (Proposition 7); from the primitive recursiveness of y = n · x, the technique of bounded quantification (Proposition 8) then insures that $\exists n \leqslant w$ (y = n · x) defines a primitive recursive relation of x, y and w; from which it follows, by 'identifying' the variables y and w (Proposition 2),[70] that $\exists n \leqslant y$ (y = n · x) defines a primitive recursive relation of x and y alone.

A reviewer of a schedule of proofs may not actually construct such a derivation, and even if he does, if he does so with no intention of preserving that derivation, he need not rewrite it (as the derivation above) so as to exhibit, in its material presentation, the achieved orderliness of its reasoning. Independently, however, of whether or not a particular reviewer actually constructs such a derivation of the formula $\exists n \leqslant y$ (y = n · x), the derivation given above does point to the following phenomena: first, a reviewer, in and as the 'working through' of another prover's schedule of proofs, engages in the situationally relevant work that makes up the 'checking' of the proofs of that schedule, and second, it is through that work that a reviewer comes to find, as his — though not idiosyncratic — achievement, both the accountable adequacy of the proofs of that schedule and, as well, simultaneously and inseparably, the accountable adequacy of the initial propositions of that schedule in supplying just the apparatus necessary for making the adequacy of the later proofs demonstrably and analytically available. In this way, the derivation

n · x	(Proposition 3)
y = z	(Proposition 5)
y = n · x	(Proposition 7)
$\exists n \leqslant w$ (y = n · x)	(Proposition 8)
$\exists n \leqslant y$ (y = n · x)	(Proposition 2)

not only comes to exhibit the accountable analyzability of the assertion that $\exists n \leqslant y$ (y = n · x) defines a primitive recursive relation, but it exhibits the adequacy of the initial propositions of the schedule (Propositions 1-10 of the schedule given earlier) in providing an apparatus for such a demonstration as well.

In the discussion that follows, I review the work of producing the formula

$x|y \Leftrightarrow \exists n \leqslant y$ (y = n · x)

as a proof of the primitive recursiveness of the divisibility relation. The aim of this review is twofold: first, it will illuminate how a prover, *in constructing a schedule of proofs*, selects and articulates the initial propositions of the schedule — and, therein, cultivates and refines locally developing practices of proving — so that the-initial-propositions/the-way-of-working-to-which-they-are-inseparably-tied come to compose, over the course of their development, an accountably adequate apparatus for the analysis of the later proofs. At the same time, this review will permit the work practices of a prover constructing a schedule of proofs to be contrasted to those of a prover reviewing another prover's schedule. Although I speak here, and in the following, of an 'original prover' and a 'reviewer,' these terms are actually being used to refer to contrasting work circumstances and styles. In consequence, the material that follows can be seen to simply open for inspection the ways in which, for prover and reviewer alike, the adequacy of the initial propositions as an analytic apparatus is itself the achievement of the local work of a schedule's production and inspection and, therein, more generally, to open for further inspection the ways in which a schedule of proofs, as a temporally developing object/organization-of-practices, provides the background for that schedule's further enunciation and organization.

To begin the review of the work of developing a proof of the primitive recursiveness of the divisibility relation, let us start, as an actual prover might, by writing the conventional definition of the divisibility relation as

$x|y \Leftrightarrow x \neq 0$ and $\exists n \, (y = n \cdot x)$

where, as the way in which that formula is written and thereby recalled, a prover will verbalize that formula — cotemporaneously with his writing it — as saying that x divides y if and only if x does not equal zero and there exists a number n such that y equals n times x.[71] In considering this formula as the initial part of a prospective and potentially realizable course of writing that, at the same time, that formula itself entails, a prover will modify or rewrite it so that y will appear as an upper bound on the existentially quantified variable n, thus:

$x|y \Leftrightarrow x \neq 0$ and $\exists n \leq y \, (y = n \cdot x)$.

That a prover will do this is integrally tied to the recognized utility of the technique of bounded quantification as a means of constructing primitive recursive relations. However, that the writing of this modified formula is tied to the recognized utility of that technique is not the phemenon of immediate interest; instead, the phenomenon of interest is the fact that the analyzability of that connection is explicitly worked out, post facto, as a thematic element of a prover's work in developing a schedule of proofs. Specifically, in coming to write $\exists n \leq y \, (y = n \cdot x)$

A SCHEDULE OF PROOFS

as replacing $\exists n \, (y = n \cdot x)$, a prover will identify, over the course of that writing, those features of the technique of bounded (existential) quantification that are of pointed relevance to establishing the applicability of that technique to the developing problem-at-hand, therein simultaneously providing for the prospective work of making analytic the assertion that $\exists n \leqslant y \, (y = n \cdot x)$ does, in fact, define a primitive recursive relation. In consequence, if a prover, at this point in his work, were to provisionally[72] formulate the technique of bounded quantification — as, for example,

if $R(x_1, \ldots, x_m, z)$, is an $(m + 1)$-place primitive recursive relation of $(x_1, \ldots, x_m, z)$, then the relation 'defined'[73] by $\exists z \leqslant y \, (R(x_1, \ldots, x_m, z))$ is a primitive recursive relation of $(x_1, \ldots, x_m, y)$

— that formulation would not be tailored[74] so as to immediately apply to the transformation of the formula $\exists n \, (y = n \cdot x)$ to $\exists n \leqslant y \, (y = n \cdot x)$, but instead, it would be tailored so that, in and as the contrast it provides between its own generality and the particular, material-specific transformation of $\exists n \, (y = n \cdot x)$ to $\exists n \leqslant y \, (y = n \cdot x)$, it provides as well for exactly those details of proving that need to be checked to insure that $\exists n \leqslant y \, (y = n \cdot x)$ does, in and of itself, accountably define a primitive recursive relation. In this way, whether or not the technique of bounded quantification is then and there explicitly formulated, the adequacy of $\exists n \leqslant y \, (y = n \cdot x)$ as defining $x|y$ as a primitive recursive relation *comes to* seeably/showably depend on the satisfaction of two conditions: (1) that $y = n \cdot x$, in itself, defines a primitive recursive relation and (2) that if $x|y$ holds between x and y, then y is, in fact, an upper bound on the quantified variable n. Concerning the later condition, a prover will differentiate and examine various 'cases' provisionally adequate to showing that y is such an upper bound and, as that examination's achievement, will establish the demonstrable adequacy of those cases for that assertion.[75]

Our review of the lived-work of producing the proof

$x|y \Leftrightarrow \exists n \leqslant y \, (y = n \cdot x)$

has taken us to the point at which a prover will have before himself the definition

$x|y \Leftrightarrow x \neq 0$ and $\exists n \leqslant y \, (y = n \cdot x)$

as a specifically-textured, developing, provisional proof of the primitive recursiveness of the divisibility relation, its provisional character being partially formulated, as the prover's achievement, by the assertion that if $y = n \cdot x$ defines a primitive recursive relation, then so will $\exists n \leqslant y \, (y = n \cdot x)$. I now want to introduce — both as a means of continuing

103

the descriptive exegesis of the work of producing the 'final' proof and as a means of contrasting a prover's and a reviewer's work in relation to that proof — an example of what, in practice, would be an unexplicated, situationally-occasioned, locally-efficacious method[76] that a prover working through another prover's schedule of proofs might employ as a means of establishing the practical adequacy of that schedule's proofs. This method, already used in introducing this topic, is this: let a displayed formula stand for the assertion that that formula defines either a primitive recursive function or a primitive recursive relation (as the case may be), and let a number enclosed in parentheses stand for the number of the proposition of that schedule of proofs that either 'justifies' that assertion or that 'justifies' it as following from the formulas that are displayed above it. Thus, a directed sequence of such formulas and numbers

A   (a)
B   (b)
.   .
.   .
.   .

G   (g)

is constructed as, and composes, as a practically accountable course of reasoning and writing, a 'derivation' establishing the last formula G of the sequence as defining (again, as the case may be) either a primitive recursive function or relation.

As one last preliminary, let us recall a statement of the technique of bounded quantification as it might appear in a schedule of proofs (as, for example, Proposition 8 of the schedule outlined earlier):

Let R be an $(m + 1)$-place numerical relation and let $\forall z \leqslant y \, (R(x_1, \ldots, x_m, z))$ and $\exists z \leqslant y \, (R(x_1, \ldots, x_m, z))$ denote the $(m + 1)$-place relations that hold for $(x_1, \ldots, x_m, y)$ if $R(x_1, \ldots x_m, z)$ holds for all $z \leqslant y$ or for some $z \leqslant y$, respectively. Then if R is a primitive recursive relation, so are $\forall z \leqslant y \, (R(x_1, \ldots, x_m, z))$ and $\exists z \leqslant y \, (R(x_1, \ldots, x_m, z))$.

Now, perhaps as the beginning of a marginal comment through which the prover intends to preserve for himself a course of reasoning adequate to the text's assertion, perhaps as the scratch paper work through which he is searching for such a course of reasoning and writing, a prover, on coming to the proposition and proof

A SCHEDULE OF PROOFS

11 The 2-place relation x|y is primitive recursive.

Proof: $x|y = \exists n \leq y \, (y = n \cdot x)$.

while working through another prover's schedule of proofs, might write

$y = n \cdot x$

$\exists n \leq y \, (y = n \cdot x)$

which can be translated as follows: if $y = n \cdot x$ defines a primitive recursive relation, then $\exists n \leq y \, (y = n \cdot x)$ will define one by construction from $y = n \cdot x$ as justified by the technique of bounded quantification.[77] One of the problems that this partial and anticipatory derivation exhibits is the problem of justifying that, on the basis of that particular schedule of proofs, $y = n \cdot x$ defines a primitive recursive relation. But another problem arises as well, whose unanticipated character resides in the fact that it is by virtue of the material specificity of the displayed equations that that problem comes to be discovered and exhibited. The problem is this: although the statement of Proposition 8 that was given above does not explicitly prohibit the use of a variable (like y) that already occurs in a formula defining a relation R (like that defined by $y = n \cdot x$) as the upper bound on the quantified variable (like n in $\exists n \leq y \, (y = n \cdot x)$), neither does that statement explicitly permit such a formal procedure.[78]

In seeing the difficulty that the notationally-specific derivation given above makes available — in seeing the problem of justifying on the basis of the statement of Proposition 8 the writing of $\exists n \leq y \, (y = n \cdot x)$ as a formula properly following the writing of $y = n \cdot x$ — a prover will (as one possibility) modify or rewrite that derivation so that it appears as

$y = n \cdot x$

$\exists n \leq w \, (y = n \cdot x)$ (8)

using w rather than y as the variable serving as an upper bound for n since this new transformation, in contrast to the one using y, is now in accord with the developed relevancies of Proposition 8 and where, parenthetically, the notational device of citing the proposition number might start to be developed as a situationally and materially motivated accounting procedure. At the same time, however, that a prover will come to write $\exists n \leq w \, (y = n \cdot x)$ as following $y = n \cdot x$, and as the directed character of that writing, a prover will have already seen that $\exists n \leq y \, (y = n \cdot x)$ results from $\exists n \leq w \, (y = n \cdot x)$ by symbolically replacing w with y, therein 'anticipating' that transformation as a prospectively accountable course of writing and reasoning. In this way, the new problem that comes to be posed, as an integral feature of the work of producing an accountable derivation, is that of justifying the writing of $\exists n \leqslant y\ (y = n \cdot x)$ as a formula properly following $\exists n \leqslant w\ (y = n \cdot x)$, a problem that we can represent as

$$y = n \cdot x \qquad (\ )$$
$$\exists n \leqslant w\ (y = n \cdot x) \qquad (8)$$
$$\exists n \leqslant y\ (y = n \cdot x) \qquad (\ )$$

Given these circumstances, a prover might thereupon interrogate the propositions of the schedule preceding Proposition 11 and come to write/reason as follows: if $\exists n \leqslant w\ (y = n \cdot x)$ defines a primitive recursive relation R in three variables — that is, if

$$R(w, y, x) \Leftrightarrow \exists n \leqslant w\ (y = n \cdot x)$$

— then

$$R(y, y, x) \Leftrightarrow \exists n \leqslant y\ (y = n \cdot x)$$

is obtained from R(w, y, x) by 'substituting' the variable y for the variable w. But Proposition 2 of the schedule (asserting that the variables in a primitive recursive function can be 'identified') seeably/showably justifies such a substitution,[79] and in consequence, the derivation that was started above can now be accountably written as[80]

$$y = n \cdot x \qquad (\ )$$
$$\exists n \leqslant w\ (y = n \cdot x) \qquad (8)$$
$$\exists n \leqslant y\ (y = n \cdot x) \qquad (2)$$

As I indicated earlier, the necessity for developing (as in the derivation above) the argument that the primitive recursiveness of the relation defined by $\exists n \leqslant y\ (y = n \cdot x)$ follows by the technique of bounded quantification from the primitive recursiveness of the relation defined by $y = n \cdot x$ potentially arises from within a reviewer's materially-specific attempt to recover the original prover's work as an accountable course of writing and reasoning. Contrastingly, given the different material circumstances of the original prover's work in constructing a schedule of proofs — summarized by pointing to his attempt *to produce* a primitive recursive definition, by pointing to his facility in writing certain formulas and, cotemporaneously, noting to himself the justification of such writing as applications of the technique of bounded quantification, and by pointing to his attempt to produce a primitive recursive definition of the divisibility relation *specifically* and, therein, to his immediate concern for, and his immediate work in, establishing the adequacy of y as an upper bound for n in the definition of x|y — the original prover 'may not give'[81] closer scrutiny to the course of combined writing and reasoning that assures that the primitive recursiveness of $\exists n \leqslant y\ (y = n \cdot x)$ accountably follows from that of $y = n \cdot x$.

A SCHEDULE OF PROOFS

Thus, without specifically reviewing the applicability of the technique of bounded quantification to that transformation, a prover constructing a schedule of proofs may come to establish the practical adequacy of the formula

$x|y \Leftrightarrow x \neq 0$ and $\exists n \leqslant y \, (y = n \cdot x)$

as a primitive recursive definition of $x|y$ contingent only on the primitive recursiveness of $y = n \cdot x$.[82]

Let us turn then to the examination of how a prover, having constructed a schedule of proofs 'so far,' comes to establish $y = n \cdot x$ as defining a primitive recursive relation.[83] The prover's problem is this: he must develop the 'earlier' parts of this-particular-schedule-of-proofs/ the-work-practices-in-which-that-schedule-is-irremediably-embedded so as to provide an apparatus/the-techniques-internal-to-that-schedule for adequately analyzing $y = n \cdot x$ as such a definition. To do so, a prover might begin by seeing, as a consequential manipulation of signs, that the equation

$y = n \cdot x$

is formed by joining

$y =$

with

$n \cdot x$

where, at the same time and intrinsic to that manipulation, the prover would recognize the equality relation and multiplication as a primitive recursive relation and function, respectively. However, in addition to the formal character of this construction, a prover will have already temporalized that construction as well — that is, $n \cdot x$ is seen as being placed 'within' the relation 'already' defined by the equality sign. A prover will then develop this temporal ordering in and as an accountable course of proving: for example, a prover might begin by writing

$K_=(x, y)$

(where $K_=(x, y)$ is intended 'to stand for'[84] the characteristic function of the equality relation) and by writing after $K_=(x, y)$

$x = y$,

thereby associating $x = y$ with $K_=(x, y)$ and, therein, providing for the projected analysis of it.[85] What a prover then wants to do is to 'substitute'[86] $n \cdot x$ for y in $K_=(x, y)$. Such a substitution, however, will result in $K_=(x, n \cdot x)$ and $x = n \cdot x$, neither of which corresponds to $y = n \cdot x$. An efficacious choice of variables[87] will remedy this difficulty: by rewriting the equation as[88]

107

A SCHEDULE OF PROOFS

$y = z \quad \quad K_=(y, z)$

the prover can then 'substitute' n · x for z so as to obtain y = n · x and $K_=(y, n \cdot x)$ and, therein, realize as well the 'temporal' character of the construction of y = n · x. Finally, evoking the functional notation f(n, x) = n · x to clarify the construction, the prover can write

$y = n \cdot x \quad \quad K_=(y, f(n, x)),$

the latter as, prospectively, the characteristic function of the relation defined by y = n · x. In that $K_=(y, f(n, x))$ can be seen as being constructed from the primitive recursive functions $K_=$ and f by substitution, $K_=(y, f(n, x))$ already seeably/showably defines a primitive recursive function, and all that remains in order to show that y = n · x defines a primitive recursive relation — as a recognized, practical requirement arising from within the way in which $K_=(y, f(n, x))$ was constructed — is to check that $K_=(y, f(n, x))$ is, in fact, the characteristic function of y = n · x. By working backward through the formulas, a prover will perceptually organize and 'rehearse' those formulas, as an accountable course of reasoning, so as to be able to read $K_=(y, f(n, x)$, as saying that if y = n · x, then $K_=(y, f(n, x)) = 1$ and if y ≠ n · x, then $K_=(y, f(n, x)) = 0$.[89] Having *in this way* insured that $K_=(y, f(n, x))$ is the characteristic function of y = n · x, the prover has demonstrated the primitive recursiveness of y = n · x by making that equation analytically available as defining such a relation from within a 'completed' course of work.

As the reader will recall, the question of the primitive recursiveness of y = n · x was addressed as the projected last step in demonstrating the primitive recursiveness of the relation defined by ∃n ≤ y (y = n · x). With the primitive recursiveness of y = n · x now established, a prover (or a reviewer) will see that he could construct (or, respectively, has found and could preserve by writing) a 'unidirectional' or 'linear' derivation of ∃n≤y (y = n · x) ad defining a primitive recursive relation.

In a short while I will give such a derivation, and I will thereby complete the derivation that I had started earlier in the discussion. However, before I do this, I want to review a slightly different way of showing that y = n · x defines a primitive recursive relation. The reason for giving this alternative construction can be developed as follows: first, in that the characteristic function of the equality relation is primitive recursive by a previous construction,[90] the construction of y = n · x given above can be formulated as, simply, the substitution of a primitive recursive function, n · x, for a variable, z, in a primitive recursive function, $K_=(y, z)$, and this construction preserves primitive recursiveness by the definition of the class of primitive recursive functions. Second, on coming to a problem similar to the one for

A SCHEDULE OF PROOFS

$y = n \cdot x$ as, for example, the problem of showing that $x \neq 0$ defines a primitive recursive relation, a prover will find, in the case of $x \neq 0$, that the solution to this problem consists 'first' — as an accountable temporal organization of work — of realizing that 0 can be written as $Z(y)$, the primitive recursive zero function, and by so writing 0, 'then' of seeing — as the result of the seeable/showable availability of a course of argumentation appropriately mimicking the one developed for $y = n \cdot x$ — that $K_{\neq}(x, Z(y))$ can be similarly 'derived' as the characteristic function of $x \neq 0$. Thus, even though the prover would find an immediate need for a construction similar to that used for $y = n \cdot x$, the increasing familiarity and naturalness of the work involved in such constructions, the (now) obviousness of those constructions as an accountable line of argumentation, and the fact that, as an extractable line of argumentation, it is already enunciated in the definition of the class of primitive recursive functions, all offer no motivation to a prover for articulating the construction of $y = n \cdot x$ as a separate technique of constructing primitive recursive relations.[91] In contrast, the method of construction $y = n \cdot x$ that I will give now does exactly the opposite; it suggests — in and as the problematic details of its own accountable production — the need for formulating just such a technique as a separate proposition of the developing schedule of proofs.

Let us suppose that a prover (or a reviewer), after seeing that the insertion of $n \cdot x$ in $y = \_$ provides a potentially accountable construction of $y = n \cdot x$, began the work of proving that $y = n \cdot x$ defines a primitive recursive relation by naming that relation T,

$T(y, n, x) \Leftrightarrow y = n \cdot x.$

Given this as the material origins of his work — and, in particular, the fact that he has associated a 'name' T with the relation defined by $y = n \cdot x$ — a prover might write next

$E(y, x) \Leftrightarrow y = z$

and

$f(n, x) = n \cdot x,$

thereby 'naming' the equality relation and multiplication E and f, respectively. Then, as the now recognized and realized directed course of his writing, the prover can construct $y = n \cdot x$ as

$T(y, n, x) \Leftrightarrow E(y, f(n, x)).$

With the formulas

$T(y, n, x) \Leftrightarrow y = n \cdot x$

$E(y, z) \Leftrightarrow y = z$

109

A SCHEDULE OF PROOFS

$f(n, x) = n \cdot x$

$T(y, n, x) \Leftrightarrow E(y, f(n, x))$

before the prover — at least as a perceptually organized course of work — the 'question' that arises is that of what to do next. More specifically, in coming to write

$T(y, n, x) \Leftrightarrow E(y, f(n, x))$,

the prover has not only 'named' the component parts of that formula, but he has given the equation $y = n \cdot x$ an incipient analytic structure. Part of the prover's search for the next thing to do consists of reviewing the construction of T to see if, in elucidating that incipient structure, that structure can be made explicit as part of a proof of the primitive recursiveness of $y = n \cdot x$.

What a prover will do next is to write

$K_T(y, n, x) = K_E(y, f(n, x))$,

and in writing this equation, and in seeing it as an appropriate thing to write at this point, a prover will come to recognize, as that equation's developingly 'self'-explicated gestalt, that what he must do now is to show, in that $K_E(y, f(n, x))$ apparently[92] equals $K_T(y, n, x)$, that $K_E(y, f(n, x))$ defines a primitive recursive function. (As part of that 'gestalt,' a prover will justify the fact that this is what he must do by reconstructing for himself — through the use of the material detail of his course of writing — the fact that a relation R is, by definition, primitive recursive if its characteristic function K_R is and, therefore, that T will be primitive recursive if K_T is.) Thus, in writing

$K_T(y, n, x) = K_E(y, f(n, x))$,

a prover has already envisioned that formula as part of this-yet-to-be-completed and, therein, conjectured course of accountable work from within which it will (prospectively) take on its character, and be established, as a naturally accountable part of a demonstration of the primitive recursiveness of $y = n \cdot x$.

Among the things that the prover has done, one of the things that he will recognize that he has done is to have formulated the original problem of showing that the relation defined by $y = n \cdot x$ is primitive recursive *as* the problem of showing that the construction of the relation T as

$T(y, n, x) \Leftrightarrow E(y, f(n, x))$

exhibits T as such a relation. Moreover, in that this realization is tied to the prover's having written (or to his having envisioned the writing of)

$K_T(y, n, x) = K_E(y, f(n, x))$,

the prover will not only undertake his ensuing work as that of finding this problem's solution, but he will have already found the material origins of that work in the formula

$$K_T(y, n, x) = K_E(y, f(n, x))$$

itself. That is, in the way in which the prover has come to formulate the problem of showing the primitive recursiveness of $y = n \cdot x$ as the problem of showing the primitive recursiveness of the relation T defined by

$$T(y, n, x) \Leftrightarrow E(y, f(n, x)),$$

the prover already has a prospective way of solving that problem through the examination of the equation

$$K_T(y, n, x) = K_E(y, f(n, x)).$$

Now, in his previously coming to write the formula

$$T(y, n, x) \Leftrightarrow E(y, f(n, x)),$$

a prover will have already identified E and f as, respectively, a primitive recursive relation and function. In consequence, the primitive recursiveness of E and f is now available to the prover as a witnessible and remarkable feature of

$$K_T(y, n, x) = K_E(y, f(n, x)).$$

In that this is so, a prover will find, in and as the organizing work of 'seeing,' that

$$K_T(y, n, x) = K_E(y, f(n, x))$$

exhibits[93] K_T as being constructed by the substitution of a primitive recursive function, $f(n, x) = n \cdot x$, for a variable, z, in the primitive recursive function $K_E(y, z)$, the characteristic function of the equality relation. Thus, in that the construction of K_T is now available as a practically accountable instance of the general process of defining a primitive recursive function by substitution, the formula

$$K_T(y, n, x) = K_E(y, f(n, x))$$

has come to seeably/showably exhibit the primitive recursiveness of K_T itself.

As the reader will recall, a prover, working to establish the primitive recursiveness of $y = n \cdot x$ in the fashion that I described first, as an endogenously articulated feature of that work, needed to check that the function $K_=(y, f(n, x))$ was the characteristic function of $y = n \cdot x$. From within the present way of working a different problem arises: Earlier in writing

$$K_T(y, n, x) = K_E(y, f(n, x))$$

A SCHEDULE OF PROOFS

beneath

$$T(y, n, x) \Leftrightarrow E(y, f(n, x)),$$

a prover would have seen in that arrangement, as the proper temporality of that arrangement, that that arrangement *can be seen* as one indicating a relation of proving — that is, that

$$K_T(y, n, x) = K_E(y, f(n, x))$$

'follows from'

$$T(y, n, x) \Leftrightarrow E(y, f(n, x)).$$

And in seeing that this was so, the prover would have also realized that he needed, as part of his current work, to establish that the one did, in fact, 'follow' from the other. So far, the prover has only shown that if

$$K_T(y, n, x) = K_E(y, f(n, x)),$$

then K_T is a primitive recursive function. Thus, given the material development of the problem itself, the problem that remains is that of showing that

$$T(y, n, x) \Leftrightarrow E(y, f(n, x))$$

and

$$K_T(y, n, x) = K_E(y, f(n, x))$$

can be embedded in a course of work such that they stand in the relationship of the latter being provable from the former.

Returning to this problem after establishing K_T as a primitive recursive function, a prover will see, as the incipient part of a solution to it, that the notation $T(y, n, x)$ can be read as (and, in fact, can now be retrospectively recalled as an abbreviation for)

$$(y, n, x) \in T,$$

i.e., that the 3-tuple (y, n, x) is a member of the relation T. In that the prover sees this, he is able to read the formula

$$T(y, n, x) \Leftrightarrow E(y, f(n, x))$$

as saying that

$$(y, n, x) \in T \Leftrightarrow (y, f(n, x)) \in E,[94]$$

therein providing for the following line of argumentation: since the characteristic function of an m-place relation R will be 1 if $(x_1, \ldots, x_m) \in R$ and will be 0 if $(x_1, \ldots, x_m) \notin R$, $K_T(y, n, x)$ will equal 1 or 0 as (y, n, x) is or is not a member of T, as $(y, f(n, x))$ is or is not a member of E, as $K_E(y, f(n, x))$ equals 1 or 0. 'Thus' — that is, as the naturally accountable consequence of this reasoning —

$K_T(y, n, x) = K_E(y, f(n, x))$,

and in that '$K_T(y, n, x)$ equals $K_E(y, f(n, x))$', as the-proof's/the-work-of-the-proof's own accomplishment in developing what needed to be shown about $y = n \cdot x$ to insure that it defines a primitive recursive relation, the demonstration that $y = n \cdot x$ defines a primitive recursive relation is thereby brought to a close.

We have now developed a second way of establishing the primitive recursiveness of $y = n \cdot x$. As I suggested earlier, this alternative way of working makes available the following contrast: from within the argument-specific work of the earlier analysis of $y = n \cdot x$, a prover will find no apparent, material motive for articulating the demonstration of the primitive recursiveness of $y = n \cdot x$ as a separate technique of constructing primitive recursive relations. In contrast, a prover working in the second manner, in coming upon the problem of showing that the construction of

$T(y, n, x) \Leftrightarrow y = n \cdot x$

as

$T(y, n, x) \Leftrightarrow E(y, f(n, x))$

preserves the primitive recursiveness of E and f, comes on that problem as a problem in its own right for which he *then* (in the sense of temporal succession) finds its solution as a *discovered* course of writing and reasoning. In consequence (but possibly a consequence realized in coming to see and work out, as a recurring argument, the accountably-same construction in subsequent parts of the schedule), a prover might formulate[95] this construction of $y = n \cdot x$ as itself an extractable course of reasoning (Proposition 7):

( ) The relation S obtained by substituting a primitive recursive function f for a variable in a primitive recursive relation R is primitive recursive.

And, by extrapolating from the particular proof for $y = n \cdot x$, a prover can obtain that proposition's proof:

Proof: Let S be the relation defined by
$S(x_1, \ldots, x_{i-1}, y_1, \ldots, y_n, x_{i+1}, \ldots, x_m) \Leftrightarrow$
$\quad R(x_1, \ldots, x_{i-1}, f(y_1, \ldots, y_n), x_{i+1}, \ldots, x_m)$.
Then[96]
$K_S(x_1, \ldots, x_{i-1}, y_1, \ldots, y_n, x_{i+1}, \ldots, x_m) =$
$\quad K_R(x_1, \ldots, x_{i-1}, f(y_1, \ldots, y_n), x_{i+1}, \ldots, x_m)$.
Hence, if R and f are primitive recursive so is S.

Finally, we should note that if the 'original' schedule of proofs did not

A SCHEDULE OF PROOFS

include this proposition, then a reviewer working in a manner similar to that just described, in coming to find the need to account for the construction of T as a primitive recursive relation, would find the faulted character of the schedule and, therein, would find as well the need to develop and add such a proposition to it.

Let us now return to the developing material proof of the primitive recursiveness of the divisibility relation,

$x|y \Leftrightarrow x \neq 0$ and $\exists n \leq y \, (y = n \cdot x)$.

In writing this equation, in and as the way in which the writing of it is embedded in the work of developing the schedule of proofs, a prover will have already noted that

$x \neq 0$ and $\exists n \leq y \, (y = n \cdot x)$

consists of '$x \neq 0$,' 'and,' and '$\exists n \leq y \, (y = n \cdot x)$' and that if '$x \neq 0$' and '$\exists n \leq y \, (y = n \cdot x)$' define primitive recursive relations separately, then so will the conjunction '$x \neq 0$ and $\exists n \leq y \, (y = n \cdot x)$.'[97] Furthermore, as we have seen above, in developing a sketch-proof that $\exists n \leq y \, (y = n \cdot x)$ defines a primitive recursive relation, a prover must 'first'[98] show that $y = n \cdot x$ defines a primitive recursive relation as well. The point to be made now is that once a construction of $y = n \cdot x$ as a primitive recursive relation has been given, an analogous construction can be given for $x \neq 0$ and, conversely, once a construction exhibiting $x \neq 0$ as defining a primitive recursive relation has been given, a similar construction for $y = n \cdot x$ follows from it: thus, by seeing that 0 can be interpreted as the zero function $Z(w) = 0$, a primitive recursive function by definition, the prover, in a fashion analogous to the construction of $y = n \cdot x$, will cotemporaneously see (as a seeably writable course of argumentation) that $x \neq 0$ can be constructed by substituting $Z(w)$ for y in $x = y$.[99] This done – and, moreover, this establishing the practical accountability of the last relevant detail of

$x|y \Leftrightarrow x \neq 0$ and $\exists n \leq y \, (y = n \cdot x)$

as the relevance of the details of that formula has itself been produced and identified over the course of the formula's construction and inspection – the prover can finally *assert*, as part of the developing schedule of proofs,

( ) The 2-place relation $x|y$ is primitive recursive.

Proof: $x|y \Leftrightarrow x \neq 0$ and $\exists n \leq y \, (y = n \cdot x)$

as the practically adequate statement and proof that the divisibility relation is primitive recursive.

Similarly, let us now return to the reviewer's examination of a proof of the primitive recursiveness of the divisibility relation as he comes to

A SCHEDULE OF PROOFS

that proof while working through another prover's schedule of proofs. For reasons that will be given shortly, the proof that will appear in a 'finished' schedule will not be the one given above, but, instead, it will most likely exclude the condition that $x \neq 0$:

$x|y \Leftrightarrow \exists n \leqslant y \, (y = n \cdot x)$.

Now, as I indicated at the beginning of this topic, a reviewer, working in the manner that I have described, in and as that work, may compose the following course of writing and reasoning as a derivation of the primitive recursiveness of $\exists n \leqslant y \, (y = n \cdot x)$:[100]

$$y = z \quad (5)$$
$$n \cdot x \quad (3)$$
$$y = n \cdot x \quad (7)$$
$$\exists n \leqslant w \, (y = n \cdot x) \quad (8)$$
$$\exists n \leqslant y \, (y = n \cdot x) \quad (2)$$

thereby not only finding and exhibiting $\exists n \leqslant y \, (y = n \cdot x)$ as defining a primitive recursive relation but finding and exhibiting, as something found and exhibited in course, the adequacy of the initial propositions of the schedule in providing such an analysis. In completing such an analysis (after having already examined the displayed formula

$x|y \Leftrightarrow \exists n \leqslant y \, (y = n \cdot x)$

as a definition of the divisibility relation), the reviewer finds, therein, as the closure of his work in reviewing that proof, the practical adequacy of the material proof in-itself as that proof appears in the schedule.

As the reader will recall, I began the discussion of this topic by using the very same analysis of the construction of $\exists n \leqslant y \, (y = n \cdot x)$ to point out that the initial propositions of a schedule of proofs (Propositions 1-10 of the schedule in which we are working) provide an apparatus for analyzing the proofs of the subsequent propositions of that schedule as defining primitive recursive functions and relations. The review of the lived-work of actually constructing and inspecting a proof of the primitive recursiveness of $x|y$ then gave us access, not to the adequacy of that apparatus in and of itself, but, instead, as real-worldly researchable phenomena, to how the practical adequacy of that apparatus is tied to the local work of constructing or reviewing such a schedule and to how that schedule-specific apparatus is itself developed from within that local work. Furthermore, the preceding discussion has indicated both how and the way in which a prover comes to articulate and, thereby, formulate as separate propositions, just the techniques that he needs to construct the later proofs of the schedule as seeably/showably defining primitive recursive functions and relations. In summary, one aspect of the local work of developing the schedule should, perhaps, be emphasized: rather than the initial propositions being constructed in a step-like, even if retrospective/prospective manner — for example, the need for a particular proposition occasioning that proposition's formulation and retrospective placement in the schedule — what the preceding material makes available is that a schedule is always and unavoidably further articulated and organized from within its development as a schedule-of-proofs/a-way-of-organizing-proving's-practices constructed 'so far.' In this way, the work of developing the schedule can be seen to be much more of a continual shaping which finds, over its course, the adequacy of, the recurring adequacy of, and the need to modify, both that schedule's material presentation and, inseparably, that schedule's developingly and endogenously articulated, proof-specific practices of proving than a construction of various elemental pieces that are arranged so as to make up the whole.

Before leaving this discussion, one last point needs to be made. A contrast has already been provided between the proof of the primitive recursiveness of the divisibility relation as that proof was obtained in the preceding discussion,

$x|y \Leftrightarrow x \neq 0$ and $\exists n \leqslant y \, (y = n \cdot x)$,

and the proof of that fact as it will more likely appear in a 'finished' schedule of proofs,

$x|y \Leftrightarrow \exists n \leqslant y \, (y = n \cdot x)$.

The exclusion of the condition $x \neq 0$ in the latter may strike the reader as curious in that, as we have seen, the definition

$x|y \Leftrightarrow x \neq 0$ and $\exists n \leqslant y \, (y = n \cdot x)$

more closely conforms with the conventional definition of the divisibility relation and, second, as we have also seen, the inclusion of the condition $x \neq 0$ does not affect the primitive recursiveness of the definition. The point is this: a prover, in consulting other, previously constructed schedules of proofs while constructing his own (or a reviewer, in working through another prover's schedule) will similarly question the absence of this condition. The inquiry that is thereby initiated finds that the exclusion of the condition $x \neq 0$ from the definition of $x|y$ really only adds one element $0|0$ to that relation. On then inspecting the subsequent propositions and proofs of the schedule that are practically available as being dependent on the definition of $x|y$ — like that of prime(x), for instance — a prover (or reviewer) will find that the inclusion of $0|0$ in the definition of division is, in fact, practically irrelevant to the schedule of proofs. In this way, then, as the provers before him, a prover will drop the restriction that x not equal 0 from his proof of the primitive recursiveness of $x|y$.

(c) *The correspondence between the propositions of a schedule of proofs and the syntax of formal number theory as an achievement of the schedule of proofs itself*

A schedule of proofs is typically organized in the following manner: an initial group of propositions (1-10 in the schedule of proofs outlined earlier) introduces those elementary primitive recursive functions and relations and those techniques of constructing primitive recursive functions and relations that are needed for the proofs of the later propositions of the schedule; a second group (11-16) then supplies an apparatus for working with the proof-specific Gödel numbering, and a final group of propositions (17-35) constructs the syntax of the particular logistic system under examination — that syntax being 'arithmetized' by the Gödel numbering — as consisting of primitive recursive functions and relations. Now the analyses of topics (a) and (b), although they pointed to, and were suggestive of, pervasive features of the lived-work of selecting and arranging the propositions of a schedule, they did this by attending particularistically to the work of producing the propositions and proofs of the first two of these three groups. The aim of the present topic is first to formulate, on behalf of the reader, a criticism concerning the applicability of those analyses to the work of developing the arithmetized syntax and, then, to reply to that criticism.

Earlier in the book I indicated that the proofs of the primitive recursiveness of the functions and relations of the arithmetized syntax were ambiguous in the following respect: one of a prover's problems in developing or reviewing a schedule of proofs is to show (or to verify) that the numerical functions and relations of the arithmetized syntax are, in fact, primitive recursive, and this amounts to showing (or verifying) that, in defining such functions and relations, those definitions provide a seeable/showable one-to-one correspondence between, for example, the elements of term(x) and the terms of P. Thus, although it is possible to provide abstractly for the collection of elements of such a relation independently of the exhibition of that collection's primitive recursiveness, in practice, at the same time that a prover is working out the demonstration of the primitive recursiveness of that collection, he is doing this by articulating the definition of an arbitrary member of it (whose arbitrariness is itself tied to the self-same demonstration). In this way, then, as lived-work, the proof of the primitive recursiveness of a component of the arithmetized syntax is the same as the primitive recursive definition of it. As a means of referring to the ambiguity of the proofs of a schedule as both proofs and definitions (an ambiguity that is formally avoided in the written presentation of the schedule), I now want to introduce and use the term 'definition/proof.'

As the reader will recall, I argued earlier that a prover's ability to write and inspect the definitions/proofs of a schedule was tied to the

A SCHEDULE OF PROOFS

prover's familiarity with the abbreviatory practices/practical techniques that are used in working with primitive recursive functions and relations. I later argued, in point 1 of the present chapter, that a Gödel numbering not only 'arithmetizes' the syntax of a formal language, but that, in doing so, it initiates — and, as a technique of proving, is itself an integral part of — the project of demonstrating that that arithmetized syntax can be defined as, and, thus, that it consists of, primitive recursive functions and relations. The notion of the 'directed' character of a schedule of proofs was then introduced to refer to this project as it unfolds in and as the work of producing a schedule.

The interrelatedness of these three themes — that the surrounding techniques of working with primitive recursive functions and relations permit the practically accountable writing and inspection of the proofs of a schedule, that a Gödel numbering is itself a technique of proving, and that a schedule of proofs has a directed character — lies in the fact that the directedness of the work of producing a schedule, as that directedness is realized over the course of developing the schedule itself, is made possible by, and is realized from within, the increasingly articulated techniques of working with primitive recursive functions, relations, and the Gödel numbering. Moreover, once the propositions and proofs concerning the arithmetized syntax have been developed and arranged, as part of their development and arrangement as part of a schedule of proofs (or once a reviewer has worked through a schedule for a first time), a prover's (or a reviewer's) cultivated familiarity with those techniques then insures that the prover (or the reviewer) can work through the schedule, *once again*, in just such an accountably sequentialized manner, at just such a pace, with just those proof-specific material details, so as to exhibit, and, therein, to realize, as a procedural way of working, the propositions concerning the arithmetized syntax *as* having been constructed in correspondence with, and by mimicking, the constructive and hierarchical definitions of the syntax of the logistic system as those definitions were specified prior to and independently of the construction of the schedule itself. In this way, then, the correspondence between the definitions of the schedule and the definitions of the 'original' syntax — as just this accountable correspondence, as a property of both the syntax and the schedule — is made available by, and is the achievement of, a schedule of proofs as a schedule-of-proofs/the-naturally-accountable-work-of-its-production.

Now the reader, by turning away from the natural accountability of a schedule *as* a local achievement, may have reasoned as follows: in that the definitions of the syntax of a formal language are[101] specified prior to the construction of a schedule, and in that the constructive and hierarchical character of those definitions demands that a prover would have already worked out a proper, sequentialized ordering of them, it should then follow that the order of the propositions of a

A SCHEDULE OF PROOFS

schedule concerning the arithmetized syntax is a necessary order available to a prover prior to the development of that arithmetized syntax; that while the proofs of those propositions still need to be constructed as materially-specific proofs, a prover need only work through a prearranged list of propositions corresponding to the previously available definitions of the syntax of the language in question, and, therefore, that the surrounding techniques of working with primitive recursive functions and relations allow the prover to work through the proofs of those propositions as a pre-formulated sequence of work. Through this reasoning, then, the reader may have seen a way of arguing that the 'directedness' of an-always-particular schedule of proofs — now interpreted as the programmatic fulfillment of a pre-established correspondence between a schedule of proofs and the syntax of a logistic system — is only dependent on the lived-work of producing that schedule as the means through which such a schedule is constructed, that that work relies on the correspondence between a schedule of proofs and the syntax of a logistic system as an objective relationship between natural, mathematical objects, and, therein, and more generally, that a schedule's discovered and demonstrable orderlinesses are, in fact, properties of a transcendental, objective proof of Gödel's theorem that the particular proof represents and to which the particular proof provides access. Thus, a reader's criticism of topics (a) and (b) may be this: that, at least for the part of a schedule of proofs concerning the syntax of the formal system in question, the orderliness of a schedule is not irremediably tied to the local practices of their discovery/exhibition, but, instead, speaks on behalf of a transcendentally existent and objective proof of Gödel's theorem.

As a means of addressing this criticism, I want to begin by bringing together several preliminary observations. These observations attend to the actual details of a schedule and the work of its production as those details provide a contrast between the syntax of a formal system and its 'representation' (via a Gödel numbering) in a schedule of proofs. By doing so, these observations offer, as a radical problem, not the problem of demonstrating that such a correspondence between syntax and schedule does not exist, but, instead, the problem of showing what such a correspondence consists of as praxis.

The first observation is this: given the practical exigencies of constructing a schedule of proofs, a prover may come to introduce certain functions and relations into the schedule that do not correspond to definitions used in specifying the original syntax of the formal language. One example from the schedule of proofs in which we are working is that of the function $n \mapsto g(x_n)$. As we shall see below, this function is introduced into the schedule partly to clarify the definition/proof of $var(x)$ that immediately follows it and partly, in and as the way in which it clarifies that definition by maintaining the sequential

A SCHEDULE OF PROOFS

character and pace of the work of working through (once again) the propositions and proofs of the schedule, to provide easier access to and to exhibit the accountable orderliness of the schedule. Another example is provided by the introduction of the function sub(x, n, a). sub(x, n, a), which specializes and modifies the arithmetized counterpart Sub(x, t, a) of the syntactically-defined operation of substitution, is introduced into the schedule specifically for its later use in the diagonalization /'proof'.

Second, in developing a schedule of proofs, a prover may find the need — again, in and as the practical exigencies of constructing that schedule — to articulate further the formal system itself by introducing syntactic definitions that were not seen to be needed for that system's original specification. An example of such an addition from within the schedule in which we are working is the definition and the proof of the primitive recursiveness of the (arithmetized) class of formation sequences of terms. Similarly, in topic (d) below, we will see that the attempt to modify the schedule of proofs in such a way that it will be applicable to extensions of our original system P necessitates[102] the definition (and the proof of the primitive recursiveness) of the function arg(x) giving the number n of terms that follow a function symbol f with Gödel number x for the concatenation of that function symbol, a left parenthesis, n terms, and a right parenthesis, in that order, to itself be a term of the language. The reader may well argue, regarding these examples, first, that the notion of a formation sequence of terms merely articulates the inductive definition[103] of the class of terms in a manner compatible with the Gödel numbering/the-techniques-of-working-with-that-numbering, and, second, that the notion of a function mapping a function symbol f to the number of arguments[104] that it takes is implicit in the definition of the function symbols as a class of primitive symbols.[105] The immediate point, however, is that the need to define the class of formation sequences of terms and the function arg(x), and the need to include those definitions in the schedule of proofs, arise from within the work of that schedule's construction.

Third, the placement and fitting of the 'new' components of the arithmetized syntax in a schedule of proofs are bound to the developing orderliness of a prover's work in constructing that schedule as the work of producing a sequentialized series of hierarchically building definitions/proofs whose construction (i.e., whose accountable work of proving) programmatically corresponds, as its achievement, to the exhibitedly-prior specification of the 'original' syntax; that is, that placement and fitting are bound, not to the transcendentalized schedule whose idealized, formal properties are merely adequate to a similarly transcendentalized proof of Gödel's theorem, but to the naturally accountable mathematical object — now understood as the pair the-material-

A SCHEDULE OF PROOFS

proof/the-practices-of-proving-to-which-that-proof-is-inseparably-tied — that the developing schedule of proofs is coming to be. Thus, although the need for a separate proposition concerning the function n $\mapsto$ g(x_n) could have been circumvented by including the proof of that function's primitive recursiveness in the definition/proof of var(x) as follows:

( ) Let var(x) be the relation that holds if and only if x is the Gödel number of a variable. var(x) is a primitive recursive relation.

Proof:[106]

$$\text{var}(x) \Leftrightarrow \exists n \leqslant x \, (x = g(x_n))$$

where g(x_n) is primitive recursive by the formula

$$g(x_n) = g((x) * \prod_{i=0}^{n} p_i^7 * g(\,))$$

thereby giving the proof of the primitive recursiveness of var(x) a nested structure, or, alternatively, the introduction of g (x_n) could have been entirely avoided by simply defining/proving-the-primitive-recursiveness-of var(x) as

$$\text{var}(x) \Leftrightarrow \exists n \leqslant x \left(x = g((x) * \prod_{i=0}^{n} p_i^7 * g(\,)) \right)$$

thereby including the formula for g(x_n) in the definition/proof of var(x) itself, these arrangements set in relief the fact that the following organization of the schedule maintains and exhibits the sequential and sequentially-paced character of the work of its proving:[107]

17 The function g(x_n) giving, for each n, the Gödel number of the n-th variable, is primitive recursive.

Proof:[108]

$$g(x_n) = g((x) * \prod_{i=0}^{n} p_i^7 * g(\,)).$$

18 Let var(x) be the relation that holds if and only if x is the Gödel number of a variable. var(x) is a primitive recursive relation.

Proof:

$$\text{var}(x) \Leftrightarrow \exists n \leqslant x \, (x = g(x_n)).$$

The inclusion and positioning in the schedule of the definition of a formation sequence of terms and the definition and proof of the primitive recursiveness of its arithmetized counterpart, formterm(x), provide a second illustration of the ways in which a prover will arrange the propositions and proofs of a schedule so as to evince the programmatic character of the correspondence between the original definitions of the syntax and the primitive recursive arithmetization of them under the proof-specific Gödel numbering. As the reader will recall from the discussion of Gödel numbering as a technique of proving, the introduction of the relation formterm(x) into the schedule — and, simultaneously, the need for explicitly defining the notion of a formation sequence of terms — was avoided by 'including' the definition of formterm(x) in the definition/proof of term(x) thus:

$$\text{term}(x) \Leftrightarrow x \neq 0 \text{ and } \exists y \leq \prod_{i=0}^{L(y)} p_i^x \; [(y)_{L(y)} = x] \text{ and}$$

$$\forall i \leq L(y) \Big([i = 0] \text{ or } [(y)_i = g(0)] \text{ or }$$
$$[\text{var}((y)_i)] \text{ or } \exists j \leq i \; \big((y)_i = g(S(\,) * (y)_j * g(\,)) \big)$$
$$\text{or } \exists j \leq i \; \exists k \leq i \; \big([(y)_i = g(+(\,) * (y)_j * (y)_k * g(\,))]$$
$$\text{or } [(y)_i = g(\cdot(\,) * (y)_j * (y)_k * g(\,))] \big) \Big)$$

In that previous discussion, in the way in which the definition/proof of term(x) was constructed — and, specifically, in that both the notion of a formation sequence of terms and the definition of sequence numbers encoding such sequences were developed so as to permit that construction — the practical adequacy of that formula for term(x) as (1) defining term(x) and (2) defining it as a primitive recursive relation was already part of the intelligibility of the formula. In the present context, however, in that the formula for term(x) has been isolated from the work of its construction, that formula loses its immediate availability as, and naturalness in, providing for and demonstrating the correspondence between the elements of term(x) and the sequences of primitive symbols defined as the terms of the formal system P.[109] Given this circumstance — or, rather, as the practical and naturally available requirements of composing a recognizably adequate schedule of proofs that this formula helps exhibit — a prover, 'instead' of giving the proof of the primitive recursiveness of term(x) as above, will more likely work up to that proof in a manner similar to the following,[110]

> Define a *formation sequence of terms* as a finite sequence of terms $t_1, \ldots, t_i, \ldots, t_m$ having the property that, for each i, $i = 1, \ldots, m$, one of the following conditions holds:
>
> (i) t_i is 0
> (ii) t_i is an individual variable
> (iii) t_i is $S(t_j)$ for some $j < i$,
> (iv) t_i is $+(t_j t_k)$ for some $j < i, k < i$
> (v) t_i is $\cdot(t_j t_k)$ for some $j < i, k < i$.
>
> ( ) formterm(x), holding if and only if x is the Gödel number of a formation sequence of terms, is a primitive recursive relation.
>
> *Proof:*
> formterm(x) $\Leftrightarrow x \neq 0$ and $x \neq 1$ and

A SCHEDULE OF PROOFS

$\forall i \leqslant L(x) \ \{(i = 0) \text{ or } ((x)_i = g(0)) \text{ or }$
$(\text{var}((x)_i)) \text{ or }$
$\exists j < i \ [(x)_i = g(S(\ )) * (x)_j * g(\ ))] \text{ or }$
$\exists j < i \ \exists k < i \ [(x)_i = g(+(\ ) * (x)_j * (x)_k * g(\ ))] \text{ or }$
$\exists j < i \ \exists k < i \ [(x)_i = g(\cdot(\ ) * (x)_j * (x)_k * g(\ ))] \ .$

( ) Let term(x) hold if and only if x is the Gödel number of a term of P. term(x) is a primitive recursive relation.

Proof:
$\text{term}(x) \Leftrightarrow \exists y \leqslant \prod_{i=0}^{L(y)} p_i^x \ \{\text{formterm}(y) \text{ and } [(y)_{L(y)} = x]\} \ .$

And, finally, as one last example of this aspect of the structuring of a schedule, let us consider the placement in a schedule of the proposition concerning sub(x, n, a). Even though sub(x, n, a) is not used in the proofs of the later propositions of a schedule, even though it is needed only for the diagonalization/'proof', and even though it does not directly correspond to a previously defined syntactic function, the definition/proof of sub(x, n, a) immediately follows that of Sub(x, t, a), the arithmetical counterpart of the syntactically defined operation of substitution, that being the recognizably proper place, as part of the produced and identifying structure of a schedule, for the work of sub(x, n, a)'s introduction.

A fourth observation is this: neither the ambient techniques of working with primitive recursive functions, relations, and a proof-specific Gödel numbering, nor the abstract correspondence that a Gödel numbering provides between the language of a formal system and the natural numbers, make obvious or automatic the ways in which the various definitions of the syntax of the formal system can, under that particular Gödel numbering, be formulated and exhibited as primitive recursive functions and relations; on the contrary, for a prover engaged in constructing the seeably/showably primitive recursive arithmetizations of the necessary syntactic objects for a schedule of proofs, the arithmetization of each such object is its own technical problem.[111]

Fifth, over the course of constructing a schedule of proofs a prover will not maintain a strict reliance on what becomes retrospectively available as the pre-existent, pre-established specification of the syntax of the logistic system in which he is working; rather, as a means of making that construction, a prover will re-examine, modify and possibly restructure the syntax itself: a prover, in working out the definitions/proofs of the arithmetized syntax, may discover various aspects of the original syntax that were not, but now need to be made explicit, like that of the notion of a variable *occurring* in a formula or of it having *no free occurrences* in a formula; a prover may come to consider whether or not commas should be included as primitive symbols or to review the way in which the functions symbols were defined and the precision required for that definition; a prover may even come to examine, in the presence of the adverse material detail of his current way of working, the efficaciousness and perspicacity of using a different notational system or a different set of axioms and rules of inference. In summary then, the specification of a formal system can be, and in fact is, continually evaluated over the course of constructing a schedule of proofs and can be and is changed if and when the need arises.[112] Moreover, that this is so points to the fact that the accountable detail of the syntax of a logistic system, for a prover, is irremediably tied to the embeddedness of those details in a structure of proving's practices and it is by recourse to that embeddedness that a prover can modify and restructure the 'original' syntax. As we shall see in the chapter 'A Structure of Proving,' a prover's ability to determine the class of formal systems for which Gödel's theorem holds becomes available to him through the inspection and analysis of a particular, materially-present proof of that theorem to find what can be modified or changed in the specification of the syntax for that particular proof to remain a proof of Gödel's theorem for that modified system.[113]

At this point, given the preceding observations, the reader may now see himself as being invited to suppose that the 'directed' character of a schedule of proofs — rendered as the sequentialized, programmatic construction of the arithmetized syntax in correspondence with a pre-established hierarchical construction of the pre-existent syntax of a logistic system — is an 'entirely' produced feature of the work of a schedule's construction and, thus, that the notion of the directedness of the lived-work of producing a schedule is only a retrospectively available attempt to provide an interior orderliness for that lived-work. One last observation must, therefore, be made: although the sequential and programmatic character of a schedule's construction, as an accountable feature of a finished schedule of proofs, is a thematic concern of a prover in constructing a schedule, although that sequential and programmatic character is specifically worked out over the course of a schedule's construction, the lived-work of that construction maintains and is informed by the directedness and the developing realization of the programmatic character of a schedule over the entire course of a schedule's construction.

Given this last observation, it would seem that we have been led to a paradox — on one hand, we have seen that the properties of a schedule of proofs are essentially tied to the local work of a schedule's production and review; on the other, it is nevertheless the case that over the course of that local work, that work retains its *sense* as the working out of an objectively and transcendentally ordered course of work

A SCHEDULE OF PROOFS

that that self-same work exhibits and to which that work provides increasingly technical access. In this way, then, these observations set in relief, and can now be reviewed in light of, the following phenomenon:[114] it is over the course of and through the work of working out the definitions/proofs of the arithmetized syntax, in and as the mutually discovered and produced compatibility of those definitions/proofs with the techniques of working with primitive recursive functions and relations, that a prover finds, as a process of rediscovery/construction, the *accountable* orderliness of the original specification of the formal syntax; that temporally developing rediscovery/construction, in the ways in which it *is* tied to the lived-work of organizing a schedule as an accountable order of proving, uncovers the orderliness of the original specification of the syntax in greater detail and as a technical, mathematical (and, therein, a mathematically analyzable) object; in consequence, in that the orderliness of the original definitions of the syntax of a formal system are re-achieved through the work of developing the schedule, the work of a schedule's construction is given its continuing sense and direction as the discovery and the working out of a prior, objective, and transcendentally-ordered course of work.

(d) *A review of the work of generalizing a schedule of proofs so as to elucidate the character of the development and organization of a schedule as a radical problem, for the prover, in the production of social order*

On various occasions in the analysis of topics (a), (b) and (c), we have been reminded that the lived-work of selecting and arranging the propositions of a schedule of proofs is, in fact, the work of re-proving a theorem that has already been proved. This circumstance can be elaborated as follows: a mathematician undertaking the construction of a proof of Gödel's theorem has himself been a prior witness to the work of that proof as an achievable organization of proving's practices; he has renewable access to the structure and detail of that proof by consulting established texts, and he uses the increasingly exhibited structure of his proof, as the exhibited structure of a proof of Gödel's theorem, to project the work that he still needs to do and to locate his immediate work within that developing, yet familiar structure. Thus, even though the innovative character of his work in bringing to material exhibition a proof of Gödel's theorem brings him into the presence of his own ability as a prover and, simultaneously, into the presence, structure, and originality of the proof itself, a prover of Gödel's theorem — in that his work is a re-proving — interprets, trivializes, projects, reviews and interrogates (as a means of further developing) his work as the work of the recovery of the familiar, remembered, inessentially modified, and proper articulation and organization of that proof's

A SCHEDULE OF PROOFS

structure and necessary detail.

In the present topic, in contrast to examining the work of proving Gödel's theorem as the work of its re-proving, I want to indicate how the orderliness of a schedule of proofs for Gödel's theorem may arise as a vital and critical production problem for a contemporary prover from within the work of proving the theorem. To do this, I begin by outlining one of the ways in which such a problem can arise and, at the same time, inseparably, by indicating what the character of that problem, as practice, is. I then proceed to work out a solution for the problem and to indicate what is required of such a solution for it to be one. The solution that is given is not original, and I have relied throughout my treatment of it on the proof of Gödel's theorem found in Mendelson's book, *Introduction to Mathematical Logic*. The point of this topic is not to give an original formulation and proof of Gödel's theorem, but to show, given the existence of already established proofs of that theorem, how the enunciation and organization of a schedule of proofs can still become a matter of critical attention and interest.

The treatment of this topic will have a technical character; it will be summarizing, and a close analysis will not be given.

In part, the material that follows has been included in order to give completeness to the discussion of the work of producing a schedule of proofs. However, in that this material illustrates how the adequacy of an established schedule can seriously be called into question from within the work of that schedule's production, and in that it illustrates how a prover will further develop the practices of proving Gödel's theorem (and, therein, further articulate the accountable structure of that proof) so as to organize and extract from those practices (as their discovered orderliness) an accountable structure of proving adequate to the solution of the problem that has been posed, the reader may gain a sense of the radical character of a prover's problem, as a problem in the local production of social order, of organizing and extracting an accountable structure of proving from within the lived-work of proving to which that structure is irremediably tied. The material in this topic will bring to a close the treatment of the lived-work of selecting and arranging the propositions of a schedule of proofs.

In order to indicate how such a problem can arise, let us begin by recalling, once again, the proof of the primitive recursiveness of term(x) as that proof was given in the discussion of Gödel numbering as a technique of proving:

$$\text{term}(x) \Leftrightarrow x \neq 0 \text{ and } \exists y \leqslant \prod_{i=0}^{L(x)} p_i^x \left\{ [(y)_{L(y)} = x] \text{ and} \right.$$

$$\forall i \leqslant L(y) \left([i = 0] \text{ or } [(y)_i = g(0)] \text{ or } [\text{var}((y)_i)] \right.$$

or $\exists j < i \, ((y)_i = g(S(\,) * (y)_j * g(\,)))$ or
$\exists j < i \, \exists k < i \, ([(y)_i = g(+(\,) * (y)_j * (y)_k * g(\,))]$ or
$[(y)_i = g(\cdot(\,) * (y)_j * (y)_k * g(\,))])\big)\bigg\}$

The reader will also recall that our development of this formula was tied to our interrelated construction of what became explicitly formulated as a 'formation sequence of terms' and the discovered compatibility of the Gödel numbers of such formation sequences with the methods of constructing primitive recursive functions and relations: to say that an expression of P is a term is to say that there exists a finite sequence of expressions of P, $t_1, \ldots, t_i, \ldots, t_m$ – i.e., a *formation sequence of terms* – such that, for each i, $i = 1, \ldots, m$, one of the following conditions holds:

(i) t_i is 0
(ii) t_i is an individual variable
(iii) t_i is $S(t_j)$ for some $j < i$
(iv) t_i is $+(t_j t_k)$ for some $j < i$, $k < i$
(v) t_i is $\cdot(t_j t_k)$ for some $j < i$, $k < i$.

Then, to each term t such as

$t = +(x_1 S(\cdot(x_5 0)))$,

we can associate a formation sequence τ, as, for example,

$\tau = x_5, \, 0, \, \cdot(x_5 0), S(\cdot(x_5 0)), x_1, +(x_1 S(\cdot(x_5 0)))$,

and, in turn, a Gödel numbering assigns a Gödel number y to such a formation sequence, as with

$y = p_1^{g(x_5)} \cdot p_2^{g(0)} \cdot p_3^{g(\cdot(x_5 0))} \cdot p_4^{g(S(\cdot(x_5 0)))} \cdot p_5^{g(x_1)}$
$\qquad \cdot p_6^{g(+(x_1 S(\cdot(x_5 0))))}$.

It is the structure of the Gödel numbering that is illustrated here that was used and articulated in the proof of the primitive recursiveness of term(x) above.

Let us now modify the proof of the primitive recursiveness of term(x) in the following way: rather than including the notion of a formation sequence of terms within the proof of the primitive recursiveness of term(x), we can, as we did in topic (c), first define the relation formterm(x) – holding if and only if x is the Gödel number of a formation sequence of terms – as a separate relation of the arithmetized syntax of P. The proof that term(x) is a primitive recursive relation then consists of two definitions/proofs:

( ) formterm(x), holding if and only if x is the Gödel number of a
    formation sequence of terms, is a primitive recursive relation.

A SCHEDULE OF PROOFS

Proof:
formterm(x) ⇔ x ≠ 0 and x ≠ 1 and

∀i ⩽ L(x)   [i ≠ 0] or [$(x)_i = g(0)$] or

[var($(x)_i$)] or

∃j < i [$(x)_i = g(S(\ ) * (x)_j * g(\ ))$] or

∃j < i ∃k < i [$(x)_i = g(+(\ ) * (x)_j * (x)_k * g(\ ))$] or

∃j < i ∃k < i [$(x)_i = g(\cdot(\ ) * (x)_j * (x)_k * g(\ ))$] } .

( ) Let term(x) hold if and only if x is the Gödel number of a term of P. term(x) is a primitive recursive relation.

Proof:
$$\text{term}(x) \Leftrightarrow \exists y \leqslant \prod_{i=0}^{L(x)} p_i^x \ \{[\text{formterm}(y)] \text{ and}$$

$$[(y)_{L(y)} = x]\} \ .$$

The reader should note that, except for the technically necessary provision that i ≠ 0, the conditions following the first bounded quantifier ∀i ⩽ L(x) in the definition/proof of formterm(x) have been arranged so as to be in serial correspondence with the five conditions used to define a formation sequence of terms and that the definition/proof of term(x) 'says' that there exists a number y such that y is the Gödel number of a formation sequence or terms $t_1, \ldots, t_m$,

$$y = p_1^{g(t_1)} \cdot \ldots \cdot p_m^{g(t_m)},$$

that t_m is a term since it has such a formation sequence, and that x is the Gödel number of the last exponent of y, that is, that x is the Gödel number of the term t_m.

Although this material restructuring the proof of the primitive recursiveness of term(x) is not critical to the argument that follows, the structure of the definition/proof of term(x) that that material restructuring exhibits is. In order to see this, let us suppose that we 'wished'[115] to generalize[116] the proof of Gödel's theorem to 'extensions' P', P'', P''', ... of our original logistic system P. An example of such an extension would be that of adding the 2-place function symbol E to the primitive symbols and of adding two nonlogical axioms,

$$E(x_1 0) = S(0)$$

and

$$E(x_1 S(x_2)) = \cdot(E(x_1 x_2)x_1),$$

to the axioms of P. The intended interpretation of E in N is exponentiation, $E(x\ y)$ being interpreted as x^y. In the section 'A Structure of Proving,' I will consider the question of whether or not the proof of

A SCHEDULE OF PROOFS

Gödel's theorem, as it has been developed in this book, can be modified so as to provide a proof of Gödel's theorem for such an extension P' of P — that is, whether or not, given the consistency (ω-consistency) of P', there exists a constructable sentence J' of P' such that neither J' (nor $\sim J'$) are deducible in P'. This is the problem of determining, given a proof of Gödel's theorem for P, the class of formal systems to which that theorem applies. In the present discussion, however, I only want to consider whether or not it is possible to modify the proof of the primitive recursiveness of term(x) so that it still holds given the addition of new function symbols to P like that of E and, hence, of new terms such as $E(x_1 x_2)$. This is obviously related to the more general problem.

If we consider just the case of the addition of the 2-place function symbol E, a solution to our problem is not hard to find and can be reconstructed as the following orderly procedure: first, an 'appropriate'[117] Gödel number needs to be assigned to the symbol E; a new clause (vi) added to the definition of a formation sequence of terms, namely

(vi) t_i is $E(x_j x_k)$ for some j < i, k < i;

and, corresponding to this clause, a new condition added to the proof of the primitive recursiveness of formterm(x), thus,

$\exists j < i \, \exists k < i \, [(x)_i = g(E(\,) * (x)_j * (x)_k * g(\,))]$.

The proof that term(x) is primitive recursive can then remain exactly as before:

$$\text{term}(x) \Leftrightarrow \exists y \leqslant \prod_{i=0}^{L(x)} p_i^x \{ [\text{formterm}(y)] \text{ and } [(y)_{L(y)} = x] \}.$$

Now, what this example makes apparent — and what, in fact, is made available not by first examining the case of the addition of one particular function symbol like E, but as the accountable structure of the reasoning of the definition/proof of term(x) that is illustrated by the case of the addition of E — is that the same procedure articulated for the function symbol E can be used for the addition of any finite number of function symbols f_r^m to the primitive symbols of P, where m indicates the number of arguments that the function symbol takes and r enumerates, and thereby distinguishes between, the m-place function symbols.[118] For this more general case, appropriate Gödel numbers again need to be assigned to each of the f_r^m — assuredly possible since only a finite number of such assignments are needed — an additional clause needs to be added to the definition of a formation sequence of terms for each of these new function symbols, and, corresponding to each such clause, a new condition needs to be added to the disjunction in the definition/proof of formterm(x),

A SCHEDULE OF PROOFS

$$\exists n_1 < i \ldots \exists n_m < i \, [(x)_i = g(f_r^m(\,) * (x)_{n_1} * \ldots * (x)_{n_m} * g(\,))].$$

The proof of the primitive recursiveness of term(x), again, remains as before,

$$\text{term}(x) \Leftrightarrow \exists y \leq \prod_{i=0}^{L(x)} p_i^x \, \{ [\text{formterm}(y)] \text{ and } [(y)_{L(y)} = x] \, \}.$$

The availability of this straightforward procedure for generalizing the proof of the primitive recursiveness of term(x) to the case where a finite number of new function symbols have been added to P provides a background for asking whether or not it is possible to prove the primitive recursiveness of term(x) for extensions of P that contain an infinite, rather than just a finite, number of function symbols. However, this question cannot be answered by further elaborating the method that we used in our previous proof: in the case of an infinite number of function symbols, our previous definition of a formation sequence of terms would have to include a new clause for each new symbol, and our previous definition/proof of formterm(x) would have to include a corresponding infinity of conditions. There is no provision in our definition or elaborations of the notion of primitive recursiveness for such a construction, and, in fact, a major theme in the historical development of the notion of primitive recursiveness was that primitive recursive functions should have finite, effective procedures for calculating their value from any given argument. Minimally, the proof of the primitive recursiveness of term(x) for such an extension of P depends on a prover being able to make the definition/proof either of formterm(x) or, in a schedule not using formterm(x), of term(x) self-referential in the sense that, for any given term t, there would be a primitive recursive procedure for deciding, through the use of the Gödel numbering and in such a way so as not to refer to an infinite list of possible constructions, whether or not the subsequences of symbols occurring in t were themselves terms. The problem is not solved by realizing that for the Gödel number of any one particular term, the number of function symbols occurring in a formation sequence for that term is finite: in that the function symbols that would need to be examined would depend on the particular term that was given, an infinite list of possible constructions from the function symbols would still be necessary in order for a procedure developed around this idea to be applicable to any possible given term.

In the material that follows, a solution to this problem is given.[119] Unlike the discussions of topics (a) and (b) however, I will not try to exhibit the lived-work of discovering and articulating the-organization-of-practices/the-materially-specific-schedule-of-proofs that makes up such a solution, although I will try to present the material in a manner that reflects the character of that solution as an (endogenous) organization of proving's work. The aims of the discussion will be these: first, the solution makes available, as its accomplishment, what a prover needs to discover, as an accountable course of reasoning, from within the lived-work of developing a schedule of proofs; thus, in comparison to the schedule of proofs outlined earlier, the solution makes available the features of that previous schedule that need to be modified or changed; and third, in these two ways, in that the solution points to the lived-work of producing a practically adequate schedule of proofs, the solution also indicates how the accountable orderliness of a schedule of proofs can become a critical production problem for a contemporary prover in and as the course of proving-again Gödel's theorem.

We wish to show that when an infinite number of new function symbols have been added to the formal system P, the terms of this new system — or, more exactly, the Gödel numbers of the terms of this new system — still constitute a primitive recursive relation. Let us begin, however, by first introducing a device that is known as course-of-values recursion:[120] As the reader will recall, a numerical function f (of one variable) is defined by primitive recursion when the value of that function at 0 is given and when the value of that function at the successor of each number x, S(x), is defined in terms of the value of the function at x — that is, if a is a natural number and h is a numerical function of two variables, then the function f defined by the equations

$f(0) = a$

$f(S(x)) = h(x, f(x))$

is said to be defined by primitive recursion. In contrast to the definition of a function by primitive recursion, a function f is said to be defined by course-of-values recursion when the value of that function at $S(x)$ depends not only on the value of f at x, but on at least some of the values $f(i)$, $i < x$.

Now, let f be a numerical function of one variable and define a function $f^{\#}$, called the course-of-values function for f, by the equation

$$f^{\#}(x) = \begin{cases} 1 \text{ if } x = 0 \\ \prod_{i=0}^{x-1} p_{i+1}^{f(i)} \text{ if } x > 0 \end{cases}.$$

The utility of this function comes from the fact that $f^{\#}(x + 1)$ 'encodes' the values of $f(i)$ for $0 \leq i \leq x$ since

$$f^{\#}(x + 1) = \prod_{i=0}^{x} p_{i+1}^{f(i)} = p_1^{f(0)} \cdot \ldots \cdot p_{x+1}^{f(x)}$$

and from the fact that these encoded values can then be recovered

A SCHEDULE OF PROOFS

from $f^{\#}$ by using the 'decoding' function $(\ )_.$, thus

$$f(i) = (f^{\#}(x+1))_{i+1}.$$

Since the course-of-values function $f^{\#}$ at $S(x)$ encodes the values of a numerical function f for all $i < S(x)$, this function's usefulness for explicit definitions by course-of-values recursion is clear. But more to the point, in order to generalize the definition/proof of term(x) to our extension of P, we will need to use two theorems concerning this function and course-of-values recursion:

Theorem 1. If $h(x, z)$ is a primitive recursive function and $f(x) = h(x, f^{\#}(x))$, then f is a primitive recursive function.
Proof: By definition,

$$f^{\#}(0) = 1$$
$$f^{\#}(S(x)) = f^{\#}(x) \cdot p_{x+1}^{h(x, f^{\#}(x))}.$$

Thus, the primitive recursiveness of $f\#$ follows from that of h by primitive recursion. Since

$$f(x) = (f^{\#}(S(x)))_{x+1},$$

where $(\ )_.$ is the 'decoding' function, f is primitive recursiveness by substitution.

Theorem 2. If $H(x, z)$ is a primitive recursive relation and a relation $R(x)$ holds exactly when $H(x, K_R^{\#}(x))$ holds, K_R being the characteristic function of R, then R is a primitive recursive relation.

In that it is not at all obvious, as of yet, how a prover might come to need and articulate the second theorem, nor, for that matter, is it obvious just what the second theorem actually proposes, I will defer its proof until later.

Let us return to the problem of proving the primitive recursiveness of term(x) when the language of P has been supplemented by the addition of an infinite number of function symbols f_r^m, where m and r are now any natural numbers greater than 0. For definiteness, let $f_1^1 := S$, $f_1^2 := +$, and $f_2^2 := \cdot$ in our new system, and let us call this new system P⁺. In previously defining/proving-the-primitive-recursiveness-of term(x) for P, we associated with each term of P a (non-uniquely determined) formation sequence for it. For example, the term $+(\cdot(x_1 0)x_2)$, by our previous method, could be associated with the formation sequence

$$\tau = x_1, 0, \cdot(x_1 0), x_2, +(\cdot(x_1 0) x_2)$$

having the Gödel number

$$p_1^{g(x_1)} \cdot p_2^{g(0)} \cdot p_3^{g(\cdot(x_1 0))} \cdot p_4^{g(x_2)} \cdot p_5^{g(+(\cdot(x_1 0) x_2))}.$$

A SCHEDULE OF PROOFS

However, we saw that our method of proving the primitive recursiveness of term(x) for P, as that method was tied to this way of associating formation sequences and Gödel numbers with the terms of P, was not conducive to the definition/proof of term(x) for P^+. In contrast, then, let us now associate a different type of formation sequence with a term $f_r^m(t_1 \ldots t_m)$ of P^+, where $t_1, \ldots, t_m$ are also terms of P^+, namely the sequence

$$f_r^m, f_r^m(, f_r^m(t_1, f_r^m(t_1 t_2, \ldots, f_r^m(t_1 t_2 \ldots t_m, f_r^m(t_1 t_2 \ldots t_m).$$

Then the term $+(\cdot(x_1 0)x_2)$, or $f_1^2(f_2^2(x_1 0)x_2)$, is now to be associated with a *unique* 'formation sequence'

$$+, +(, +(\cdot(x_1 0), +(\cdot(x_1 0) x_2, +(\cdot(x_1 0) x_2)$$

which, in turn, has the Gödel number

$$p_1^{g(+)} \cdot p_2^{g(+()} \cdot p_3^{g(+(\cdot(x_1 0))} \cdot p_4^{g(+(\cdot(x_1 0)x_2)} \cdot p_5^{g(+(\cdot(x_1 0)x_2))}.$$

Next, let us define a relation f-symbol(x) which holds if and only if x is the Gödel number of a function symbol of P^+ and a function arg(x),

$$\arg(x) = \begin{cases} m \text{ if } x = g(f_r^m) \\ 0 \text{ otherwise} \end{cases},$$

which gives the number m of terms that properly follow a given function symbol f_r^m for the concatenation of that function symbol, a left parenthesis, m terms, and a right parenthesis, in that order, to itself be a term of P^+. I will leave for later the problem of arranging the Gödel numbering in such a way so as to permit the demonstration that f-symbol(x) and arg(x) are primitive recursive. With these definitions in hand, then, we can articulate the methodic character of our new way of associating formation sequences, and thereby Gödel numbers, to the terms of P^+ as the following, provisional definition/proof of term(x) as a primitive recursive relation, where the contingencies of that definition/proof are themselves locally identified and cultivated features of that formula's construction:

term(x) $\Leftrightarrow$ x = g(θ) or var(x) or

$\exists y \leqslant \Box$ $\{[y \neq 0]$ and $[x = (y)_{L(y)}]$ and $[\text{f-symbol}((y)_1)]$

and $[L(y) = \arg((y)_1) + 3]$ and $[(y)_2 = (y)_1 * g(()]$ and

$\forall u \leqslant L(y) ([(2 < u)$ and $(u \leqslant \arg((y)_1) + 2)]$ imply

$\exists v \leqslant x$ (term(v) and $(y)_u = (y)_{u \dot{-} 1} * v$)) and

$[(y)_{L(y)} = (y)_{L(y) \dot{-} 1} * g())] \}$.

The check that $\prod_{i=0}^{L(x)+3} p_i^x$ is an upper bound for y is left to the reader.

133

A SCHEDULE OF PROOFS

An immediate problem with this definition/proof of term(x) is that it is 'impredicative' — the presence of term(v) in the definition of term(x) introduces a circularity into the definition. But by considering our new method of associating with a term t_i of P^+ a unique formation sequence and, thereby, the unique Gödel number n of that formation sequence, a prover can see/show that if a function symbol f_r^m is to appear in t_i, then the Gödel number of f_r^m must be less than n. Thus, only a finite number of, in principle, determinable function symbols need to be examined as being used in forming t_i. Even more importantly, however, we can see from the new formation sequence

$+, +(, +(\cdot(x_1\, 0), +(\cdot(x_1\, 0)\, x_2, +(\cdot(x_1\, 0)\, x_2$

for $+(\cdot(x_1 0)x_2)$ that each of the successive elements of the sequence consists of an addition of symbols to the immediately preceding element of that sequence and, therefore, that the Gödel number of each succeeding element of that sequence is greater than that of all those elements that have preceded it and that the Gödel number of a term — i.e., of the last element of such a construction sequence — has the largest Gödel number of all the elements in that sequence. Since, then, all the Gödel numbers needed to show that some number x is the Gödel number of a term are less than x, $K_{term}(x)$, the characteristic function of the terms of P^+, depends only on the values of $K_{term}(y)$ for $y < x$, and the apparent impredicativity of the preceding definition seems to be avoidable through the use of the course-of-values recursion. More precisely, we want to show that term(x) is a primitive recursive relation or, equivalently, that K_{term} is a primitive recursive function or that $\{ x \mid K_{term}(x) = 1 \}$ is a primitive recursive relation. Let us replace term(v) in our provisional definition of term(x)

term(x) ≡ x = g(0) or var(x) or
$\exists y \leq \prod_{i=0}^{L(x)+3} p_i^x \; \{[y \neq 0]$ and $[x = (y)_{L(y)}]$ and
[f-symbol$((y)_1)$] and $[L(y) = \text{arg}((y)_1) + 3]$ and
$[(y)_2 = (y)_1 * g(\,)]$ and
$\forall u \leq L(y) \,([(2 < u$ and $u \leq \text{arg}((y)_1) + 2)]$ imply
$\exists v \leq x$ (term(v) and $(y)_u = (y)_{u \dot{-} 1} * v))$ and
$[(y)_{L(y)} = (y)_{L(y) \dot{-} 1} * g(\,))]\}$

with $(z)_{v+1} = 1$ and define a new relation H(x, z),

H(x, z) ⇔ x = g(0) or var(x) or $\exists y \leq \prod_{i=0}^{L(x)+3} p_i^x \; \{[y \neq 0]$ and
$[x = (y)_{L(y)}]$ and [f-symbol$((y)_1)$] and $[L(y) = \text{arg}((y)_1) + 3]$

and $\forall u \leqslant L(y)$ $([(2 < u)$ and $u \leqslant \arg((y)_1) + 2)]$ imply
$\exists v \leqslant x$ $((z)_{v+1} = 1$ and $(y)_u = (y)_{u \dot{-} 1} * v))$
and $[(y)_{L(y)} = (y)_{L(y) \dot{-} 1} * g(\ ))] \}$.

Since $(K_{term}{}^\#(x))_{v+1} = 1$ implies, by the definition of the course-of-values function, that $K_{term}(v) = 1$, it follows, by definition of a characteristic function, that v is the Gödel number of a term and that x is the Gödel number of a term if and only if $H(x, K_{term}{}^\#(x))$. Thus, term(x) is defined by

term(x) $\Leftrightarrow H(x, K_{term}{}^\#(x))$,

and we only need to show that $K_{term}(x) = K_H(x, K_{term}{}^\#(x))$ is a primitive recursive function. But as the reader will recall, Theorem 1 stated earlier says that if $K_H(x, z)$ is a primitive recursive function and $K_{term}(x) = K_H(x, K_{term}{}^\#(x))$, then $K_{term}(x)$ is a primitive recursive function as well. Since, subject to the demonstration of the primitive recursiveness of f-symbol(x) and arg(x), H(x, z) is a primitive recursive relation, K_H is a primitive recursive function, and we are done. Moreover, if we rewrite this last argument, we see that that argument is a proof of a particular case of our second theorem concerning course-of-values recursive (with $R(x) = K_{term}(x)$):

Theorem 1. If $h(x, z)$ is a primitive recursive function and $f(x) = h(x, f^\#(x))$, then f is a primitive recursive function.

Theorem 2. If $H(x, z)$ is a primitive recursive relation and a relation $R(x)$ holds exactly when $H(x, K_R{}^\#(x))$ holds, K_R being the characteristic function of R, then R is a primitive recursive relation.

Proof: Since R(x) holds if and only if $H(x, K_R{}^\#(x))$,

$K_R(x) = K_H(x, K_R{}^\#(x))$,

and it follows from Theorem 1 that K_R is a primitive recursive function, and, hence, that R is a primitive recursive relation.

To finally complete the proof that term(x) is a primitive recursive relation for P^+, we need to show that f-symbol(x) and arg(x) are, respectively, a primitive recursive relation and a primitive recursive function. I will not belabor the technical details involved in working out these proofs.[121] 'First,' the Gödel numbering must be re-specified so as to accommodate the infinite number of function symbols. This may be done by defining

$g(f_r^m) = 9 + 8\,(2^m \cdot 3^r)$

and arranging the other Gödel numbers so that this is possible. The 'appropriateness' of this numbering can 'then' be seen to consist of the following: first, the introduction of the primitive recursive function qt(x, y), the quotient of y divided by x, allows arg(x) to be defined as the primitive recursive function

arg(x) = (qt(8, x $\dot{-}$ 9))$_1$;

second, the numbering $g(f_r^m) = 9 + 8\,(2^m \cdot 3^r)$ affords an easy proof that f-symbol(x) is a primitive recursive relation, namely

f-symbol(x) ⇔ ∃m ⩽ x ∃r ⩽ x (x = 9 + 8 $(2^m \cdot 3^r)$);

and, third, this numbering provides a method for generalizing the proof of Gödel's theorem to logistic systems that extend P through the introduction of an infinite number of constant symbols a_r (with $a_1 := 0$) and an infinite number of relation symbols A_r^m (with $A_1^2 :=$) as well as an infinite number of function symbols: for example, we can further modify our Gödel numbering so as to permit the definitions

$g(a_r) = 7 + 8r$ 1 ⩽ r

$g(f_r^m) = 9 + 8\,(2^m \cdot 3^r)$ 1 ⩽ m, 1 ⩽ r

$g(A_r^m) = 11 + 8\,(2^m \cdot 3^r)$ 1 ⩽ m, 1 ⩽ r

and by doing so, thereby insure that the Gödel numbers of the primitive symbols of the extended system are all odd and distinct, that the relations constant(x), f-symbol(x), and R-symbol(x) of the arithmetized syntax of that system are demonstrably primitive recursive, and that $g(f_r^m)$ and $g(A_r^m)$ can be 'decoded' so as to determine, in a primitive recursive manner, the number of arguments that f_r^m and A_r^m take, and all these as relevant features of the Gödel numbering that are discovered and developed over the course of, and in and as, the materially-specific work of producing a schedule of proofs for a proof of Gödel's theorem.

With these last details concerning the assignment of Gödel numbers, we have, in a sense, completed the definition/proof of term(x) for an extension P⁺ of the system P. In a finished schedule of proofs, however, the reasoning indicated in the discussion above would be arranged, not in the fashion of the preceding discussion, but as an orderly course of proving's practices providing, in and as the accountably-exhibited work of proving-again Gödel's theorem, a schedule of proofs for that proof. Moreover, that ordering of the schedule would not simply be retrospectively imposed on the lived-work of the discovery and articulation of the definition/proof of term(x), but, instead, that arrangement of practices would itself arise from within, and, in doing so, would already be tied to, the components of that definition/proof as the components of a temporally discovered, temporally developing, and temporally articulated gestalt of practices needed for, and accountably making up, the proof of the primitive recursiveness of term(x): thus, the proposition that qt(x, y) is a primitive recursive function and the

A SCHEDULE OF PROOFS

propositions needed to give such a proof would be placed in the earlier part of a schedule of proofs; the introduction of the techniques of course-of-values recursion would probably follow the techniques for working with primitive recursive functions and relations; the Gödel numbering might be specified just prior to the propositions concerning the apparatus for working with that numbering, and the positioning of the propositions concerning arg(x), f-symbol(x), term(x) and the deletion or modification of formterm(x) would all be placed in accountably appropriate and practically efficacious places in the latter part of the schedule. And through this arrangement as well, as the reader will see in the next chapter 'A Structure of Proving,' the accountable structure of the schedule of proofs as a schedule of proofs for a proof of Gödel's theorem is thereby both, and at the same time, further articulated and maintained throughout the lived-work of producing a proof of the primitive recursiveness of term(x) for P^+.

Let me bring this material to a conclusion.

In this topic, I have tried to indicate the reworking of our original schedule of proofs that is necessary to prove the primitive recursiveness of term(x) for an extension P^+ of our original language P. As the reader will recall, the discussion of topic (d) began with the observation that the work of proving Gödel's theorem now has, for a contemporary prover, the character of being the re-proving of an established theorem. What the review of the modification of the definition/proof of term(x) for P^+ has shown is that the accountable orderliness of a schedule of proofs for a proof of Gödel's theorem can still arise, for a prover proving-again Gödel's theorem, as an open problem whose solution is not guaranteed in the sense that it is not available as an extractable course of materially-exhibited reasoning from a previously established schedule of proofs and that, therefore, the identifying orderliness of a schedule of proofs can still arise for a prover as a critical and vital production problem. Thus, the preceding discussion has opened up another problem: in searching for the extension of the definition/proof of term(x), in the way in which the accountable orderliness of our schedule of proofs was thereby re-opened as a production problem, the problem of descriptively analyzing what constitutes the identifying orderliness of the lived-work of producing a schedule of proofs for a proof of Gödel's theorem — as that identifying orderliness is made available from within that work itself — becomes available as a radical problem in the study of the production of social order. I will return to this matter in the next chapter 'A Structure of Proving.'

4 *The work of providing a consistent notation for a schedule of proofs articulates that schedule as one coherent object*[122]

For brevity, I will refer in this section to the work of a prover in

A SCHEDULE OF PROOFS

developing and using a schedule of proofs' always proof-specific notation as 'notation's work.' A review of the material in the preceding section shows, in addition to that work's omnipresence and diversity, its occasioned and practical character, its unremittingly circumstantial specificity, and, through the exhibition of some of its inspectable detail, that work's utter familiarity to and unremarkableness for the experienced prover. In this section I bring to a close the discussion of the work of producing a schedule of proofs by attempting to find, in notation's work, a distinctive order-productive phenomenon. The section begins by recalling the definition of the function Sub(x, t, a). Although the naturally accountable achievement of that definition is the provision for a function, designated as Sub(x, t, a), whose properties are independent of its particular notational representation, I argue that not only is such an achievement made available through a prover's locally developing notational apparatus, but it is the embeddedness of the objects of a schedule of proofs in notation's work that provides both for the analyzability of those objects and for their utility for developing that schedule. As a first step in my argument, I describe, in a naturalistic manner, some of the accountable detail of the notational designation Sub(x, t, a). This material provides (what will later become available as) a background of notational practices which, in turn, will enable us to examine a particular instance of notation's work. A provisional analysis of that work will then bring the discussion of this section to a close.

To begin, then, let us recall the definition of Sub(x, t, a):

$$\text{Sub}(x, t, a) = \begin{cases} g(S^x_t \, A|) & \text{if } x = g(x), \, t = g(t), \, a = g(A), \\ & x \text{ an individual variable, } t \text{ a} \\ & \text{term, and } A \text{ a wff} \\ a & \text{otherwise} \end{cases}$$

where $S^x_t \, A|$ denotes the wff that results when a term t is substituted for the free occurrences of an individual variable x in a wff A.

Now, although the symbols x, t and a make this definition notationally specific, the definition of Sub(x, t, a) is constructed (and necessarily so for it to be a naturally accountable mathematical definition) so as to provide for x, t, and a only as a scheme of reference — no properties either of the symbols x, t, or a or of any 'meanings' associated with them are exhibited as necessary features of that definition. In this way, the definition of the numerical function denoted as Sub(x, t, a) provides for that function as an 'ideal' (or 'real' or 'transcendentalized') object disengaged from the local work circumstances of its development and use. If, however, we modify our definition of Sub(x, t, a) by replacing x, t and a with a, y, and x, respectively, we obtain

A SCHEDULE OF PROOFS

$$\text{Sub}(a, y, x) = \begin{cases} g(S_y^a\ X|) & \text{if } a = g(a),\ y = g(y),\ x = g(X), \\ & a \text{ an individual variable, } y \text{ a} \\ & \text{term, and } X \text{ a wff} \\ x & \text{otherwise} \end{cases}$$

which, though not 'wrong,' begins to indicate an (as-of-yet-unidentified-as-practice) confusion. We can, of course, go even further: by removing the alphabetic coordination between the numerical and syntactic variables, we can write the definition of Sub(w, y, x) as

$$\text{Sub}(w, y, x) = \begin{cases} g(S_x^a\ W|) & \text{if } w = g(a),\ y = g(x),\ x = g(W), \\ & a \text{ an individual variable, } x \text{ a} \\ & \text{term, and } W \text{ a wff} \\ x & \text{otherwise} \end{cases}$$

thereby obtaining a definition that seriously obscures the object that that definition offers for consideration.

In the presence of such possibilities, a mathematician might well argue that while a prover does, in fact, develop and use particular notational devices in constructing such a definition, he does so as a means (though, perhaps, a necessary one) for exhibiting the *independence* of that definition from what that definition makes available, as its achievement, as simply a convenient choice of notational designations for it. One of the aims of this section is to begin to find what the lived-work of such an achievement actually consists of.

Before attempting this, and as a means of providing background material for that attempt, I want to first review some of the accountable details of the notation Sub(x, t, a) as those details provide for the practical accountability and efficacious use of Sub(x, t, a) as part of a schedule of proofs. One such detail, for example, is tied to the fact that the definition of the syntactic operation of substitution — defined as the operation of replacing the free occurrences of an individual variable x in a wff A by a term t and denoted as $S_t^x\ A|$ — although it is itself defined in such a way that its variables serve as a scheme of replacement, uses its syntactic variables x, t, and A in a manner that is coordinated throughout the development and discussion of the syntax of P — namely, that x is understood as representing an arbitrary individual variable, that t represents an arbitrary term, and that A represents an arbitrary wff. The point, as it regards Sub(x, t, a), is this: the designation of the arithmetized counterpart of the substitution operation as Sub(x, t, a) used the (never made explicit) alphabetic pairing of the variable x with the Gödel number of an individual variable x, the variable t with the Gödel number of a term t, and the variable a with the Gödel number of a wff A

139

A SCHEDULE OF PROOFS

x ⟷ *x*

t ⟷ *t*

a ⟷ *A*

as a local[123] device for recalling the 'meaning' or 'interpretation' of Sub(x, t, a) in terms of $S_t^x \, A \,|$.

A second accountable feature of the notational designation Sub(x, t, a) that I wish to call to the reader's attention is that the arguments of the function so designated are arranged in a particular order — that x is the first variable, that t is the second, and that a is the third — and, thereby, as a developing aspect of the reliance of a prover on the association of x with *x*, t with *t*, and a with *A*, that if the numerical value m_1 of the first variable is the Gödel number of an individual variable, if the numerical value m_2 of the second variable is the Gödel number of a term, and if the numerical value m_3 of the third variable is the Gödel number of a wff, then the value of Sub(m_1, m_2, m_3)[124] is the Gödel number of the wff that results when the term with Gödel number m_2 is substituted for the free occurrences of the variable with Gödel number m_1 in the formula with Gödel number m_3. Although the ordering of these variables is theoretically arbitrary — for example, the variables could be arranged as in Sub(a, t, x) but still correspond to $S_t^x \, A \,|$ — that ordering is tied to a way of remembering the association of the particular variables in Sub(x, t, a) (independently of, in addition to, and in the absence of, the mnemonic use of the alphabet) with the particular variables in $S_t^x \, A \,|$, thus,

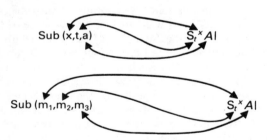

Specifically, this arrangement of variables is tied both to the temporal organization of the writing of $S_t^x \, A \,|$ as

$$S \to S^x \to S_t^x \to S_t^x \, A \to S_t^x \, A \,|$$

and, thereby, to a way of seeing the symbols $S_t^x \, A \,|$ as a temporally organized configuration,[125] and, as well, to the facility of constructing the function

$$\varphi(u) = \text{sub}(g(x_2), u, u) = g(S_{k(u)}^{x_2} \, U \,|)$$

in the diagonalization/'proof' by fixing the *first* variable in sub(x, n, a) as opposed to fixing the second variable in a function sub(n, x, a) and writing

$$\varphi(u) = \text{sub}(u, g(x_2), u),$$

a formula that would result from writing Sub(t, x, a), instead of Sub(x, t, a), as the arithemetized counterpart to $S_x^t\ A|$, instead of $S_t^x\ A|$, where, reversing the placement of t and x, $S_x^t\ A|$ now stands for the operation of substituting a term t for the free occurrences of x in A.[126]

Finally, as a means of pointing to a third accountable detail of 'Sub(x, t, a)', I want to contrast the notation for Sub(x, t, a) with that for the function sub(x, n, a) defined by the equation

$$\text{sub}(x, n, a) = \begin{cases} g(S_{k(n)}^x\ A|) & \text{if } x = g(x), a = g(A), \\ & x \text{ is an individual variable,} \\ & \text{and } A \text{ is a wff} \\ a & \text{otherwise.} \end{cases}$$

These equations define sub(x, n, a) as Sub(x, g(k(n)), a) where k(n) = $\overbrace{S(\ldots S(0) \ldots)}^{n}$, the numeral representing the number n in P. As with all the components of the arithmetized syntax, Sub(x, t, a) and sub(x, n, a) use an abbreviation of the syntactic operations or relations that correspond to them as a mnemonic device for associating the arithmetized syntax with the original syntax of the formal language. In the case of Sub(x, t, a) and sub(x, n, a), however, both of these functions correspond to the operation of substitution, and since they cannot be accountably distinguished through the alphabetic designation of their variables (both functions simply being defined on N^3), then, again as a local and locally transparent device, the two functions are distinguished by capitalizing the first letter of one of them.

Later in this section, I will return to, and reconsider, what I have spoken of as the accountable 'details' of a notational designation — in this case, of Sub(x, t, a) — as a residue of a prover's notational practice. In fact, the immediate utility of the preceding material is that it provides a background of such practices against which we can examine a prover's work in developing a schedule of proofs' notation.

In outlining a schedule of proofs at the opening of this chapter, and throughout this book as a whole, I have been following closely the proof of Gödel's theorem that is found in Robbin's *Mathematical Logic: A First Course.*[127] In the schedule of proofs given earlier, however, instead of adopting Robbin's notation for the relations occur(a_1, x, a_2, a), bound(a_1, x, a_2, a), and free(a_1, x, a_2, a), I used w, x, y, and z as the variables for these relations, writing them as occur(w, x, y, z), bound(w, x, y, a), and free(w, x, y, a) in Propositions

A SCHEDULE OF PROOFS

23, 24 and 25, respectively. The point of my doing so at that time, as a revision of Robbin's schedule considered in detachment from the work of its proving, was to emphasize that those relations were defined on N^4, specifically disengaging their definitions from the (informal, unexplicated) alphabetic correspondence that was being used in Robbin's proofs.[128] Now, there is nothing critically wrong with the notation that I adopted; in fact, the difference between Robbin's and my notation is accountably inessential to the definition/proofs of those relations. Nevertheless, from within the lived-work of developing or reviewing a schedule of proofs, Robbin's notation exhibits its preferred character. We will begin to have descriptive access to notation's work by addressing the problem of determining the local motives — in and as the work of producing/recovering the schedule of proofs as a naturally accountable course of proving — for so preferring Robbin's notation.

First, consider the definition/proof of the relation occur using my notation:

occur(w, x, y, z) $\Leftrightarrow$ wff(z) and var(x) and z = w * x * y.

What this definition 'says' is that occur(w, x, y, a) holds if and only if z corresponds, under the Gödel numbering, to a wff Z, w corresponds to an initial string W of primitive symbols of Z, x corresponds to an individual variable x, y corresponds to a terminal string Y of primitive symbols of Z, and Z can be written as the concatenation of the symbols of W, x, and Y in that order, $Z = WxY$.

At first, the definition/proof of occur(a_1, x, a_2, a) in Robbin's notation may seem only a little different:

occur(a_1, x, a_2, a) $\Leftrightarrow$ wff(a) and var(x) and a = a_1 * x * a_2.

In contrast to the earlier one, this last definition maintains the use of a as the Gödel number of a wff to be designated as A and maintains,[129] as well, the use of x as the Gödel number assigned to an individual variable to be designated as x. However, the more significant difference is that, rather than using an arrangedly arbitrary assignment of the letters w and y to represent the Gödel numbers of initial and terminal sequences W and Y of primitive symbols in the wff Z corresponding to the number z, $Z = WxY$, the definition/proof of occur(a_1, x, a_2, a) uses the designations a_1 and a_2 as a means of providing a further correspondence between A and its decomposition as $A = A_1xA_2$. This association of a_1, x, a_2 and a with $A = A_1xA_2$ is the achievement of the-proof's-notation/that-notation's-associated-work and is nowhere, except in and as that work, made explicit.

The differences between these two definitions/proofs becomes clearer in the subsequent definition/proof of bound(a_1, x, a_2, a). Roughly, bound(a_1, x, a_2, a) asserts that the occurrence of x in A indicated as A_1xA_2 is a bound occurrence of that variable — that is,

A SCHEDULE OF PROOFS

that that x occurs in a well-formed part $\forall xC$ where the exhibited occurrence of x in $\forall xC$ may or may not be the bound occurrence of x indicated in bound(a_1, x, a_2, a). Robbin's definition/proof of this relation is

bound(a_1, x, a_2, a) $\Leftrightarrow$ occur(a_1, x, a_2, a) and

$\exists c_1 \leq a \; \exists c_2 \leq a \; \exists c \leq a$ (wff(c) and $L(c_1) \leq L(a_1)$)

and $L(c_2) \leq L(a_2)$ and $a = c_1 * g(\forall) * x * c * c_2$).

If we compare this definition/proof to one using the notation bound(w, x, y, z),

bound(w, x, y, z) $\Leftrightarrow$ occur(w, x, y, a) and

$\exists u \leq z \; \exists v \leq z \; \exists p \leq z$ (wff(p) and $L(u) \leq L(a_1)$)

and $L(v) \leq L(y)$ and $z = u * g(\forall) * x * p * v$),

we see, first, that in Robbin's proof the numerals 1 and 2 are used to develop and further articulate the decomposition of a formula A associated with the number a into distinguished sequences of symbols — c_1 indicating an initial sequence C_1 of A, c_2 a terminal sequence C_2 of A, and c, with no subscript, a wff C 'between' C_1 and C_2 — with the formula A being decomposed into possible sequences of primitive symbols $A = C_1 \forall x C C_2$. Although the relationship between c_1, c_2, and c is slightly different from that between a_1, a_2, and a, the similarity between the use of these letters and subscripts is recognized by a prover in and as its practical consequentiality for finding and exhibiting the naturally accountable definition/proof of bound(a_1, x, a_2, a); those similarities, as well as the difference in their use, are witnessed in and as the work that that notation 'performs' in the proof. But while a prover depends on such notational practices, those practices need not be specified nor is it immediate, in each particular case, how that use would be specified.

In contrast to Robbin's definition/proof of bound(a_1, x, a_2, a), the adequacy of the second proof is not as immediate as Robbin's; a prover is required to do more work to find it. Once the achievement of Robbin's definition/proof is available to a prover, however, the prover is aided in finding the adequacy of the definition/proof of occur(w, x, y, z) by the analogy between that definition/proof and the earlier one. In both cases, as a prover's achievement, the work that the notation of those proofs does in making the adequacy of those proofs available is hidden in and as the natural accountability of the proofs themselves.

Given this circumstance, an additional, more complex example may help illustrate the work that Robbin's notation performs in finding and exhibiting the natural accountable definitions/proofs of his schedule.

A SCHEDULE OF PROOFS

To this end, consider the problem of proving the primitive recursiveness of the relation freefor(t, x, a) (Proposition 31) which is defined as holding if and only if there are no free occurrences of an individual variable x (corresponding, under the Gödel numbering, to the number x) in a well-formed part $\forall yc$ of the wff A (corresponding to the number a) where y is a variable occurring in the term t (corresponding to the number t). The idea of this definition is that if freefor(t, x, a) holds, then, in substituting t for x in A, a variable y in the term t will not occur within the scope of a universal quantifier in y which would make that occurrence of y a bound occurrence.

For the immediate purposes of proving the primitive recursiveness of freefor(t, x, a) — specifically, by providing the necessary level of detail to provide for that proof's associated work — let us first introduce the relation free(a_1, x, a_2, a):

free(a_1, x, a_2, a) $\Leftrightarrow$ occur(a_1, x, a_2, a) and not-bound (a_1, x, a_2, a).

free(a_1, x, a_2, a) holds if and only if the occurrence of x in A indicated by $A_1 x A_2$ is not a bound occurrence and, as a consequence, that a term t can be substituted for x at that place.

With this background, Robbin then gives the now somewhat transparent definition/proof of freefor(t, x, a), further exhibiting and developing his notational practices in doing so:[130]

freefor(t, x, a) $\Leftrightarrow$ term(t) and var(x) and wff(a) and

$\exists a_1 \leqslant a\ \exists a_2 \leqslant a$ {free(a_1, x, a_2, a) implies

  not-$\exists c_1 \leqslant a\ \exists c_2 \leqslant a\ \exists c \leqslant a\ \exists y \leqslant a\ \exists t_1 \leqslant t$

  $\exists t_2 \leqslant t$ [wff(c) and var(y) and L(c_1) $\leqslant$ L(a_1)

    and L(c_2) $\leqslant$ L(a_2) and t = t_1 * y * t_2 and

    a = c_1 * g($\forall$) * y * c * c_2]} .

With such an organization of notational practices, the associated work of the material display through which the natural accountability of that proof is achieved is readily made available and exhibited; without such systematic use of notation, the proof of the primitive recursiveness of freefor(t, x, a) becomes extremely complex, as the following definition/proof of it, using a 'more arbitrary'[131] notation, illustrates:

freefor(x, y, z) $\Leftrightarrow$ term(x) and var(y) and wff(z) and

$\exists u \leqslant z\ \exists y \leqslant z$ {free(u, y, v, z) implies

  not-$\exists r \leqslant z\ \exists s \leqslant z\ \exists p \leqslant z\ \exists q \leqslant z\ \exists b \leqslant t\ \exists c \leqslant t$

  [wff(p) and var(q) and L(r) $\leqslant$ L(u) and L(s) $\leqslant$ L(v) and x =

    b * q * c and z = r * g($\forall$) * q * p * s]} .

A SCHEDULE OF PROOFS

Given the need for such a number of numerical variables in this proof, along with the obscurity of the work required to define/prove freefor(x, y, z), a prover would find the need to begin to systematize his notation and to introduce that orderliness into his schedule of proofs.

With this material in hand, it is now possible to begin to find in a schedule of proofs' notation, as that schedule's notation is tied to its lived-work, a distinctive order-productive phenomenon in the study of mathematicians' work.

Already in this chapter we have seen that a schedule of proofs is a curiously paired object the-material-schedule/the-practices-of-proving-to-which-that-schedule-is-irremediably-tied. To speak of the naturally accountable schedule of proofs is to speak of that paired object. And, similarly, to speak of the naturally accountable definition/proof of the relation freefor(t, x, a) is to speak of the intrinsically tied pair the-material-proof/the-practices-of-proving-to-which-that-proof-is-irremediably-tied. In this way then, the 'material proof'

freefor(t, x, a) ⇔ term(t) and var(x) and wff(a) and

$\exists a_1 \leqslant a \ \exists a_2 \leqslant a \ \{\text{free}(a_1, x, a_2, a)$ implies

not-$\exists c_1 \leqslant a \ \exists c_2 \leqslant a \exists c \leqslant a \ \exists y \leqslant a \ \exists t_1 \leqslant t$

$\exists t_2 \leqslant t \ [\text{wff}(c)$ and var(y) and $L(c_1) \leqslant L(a_1)$

and $L(c_2) \leqslant L(a_2)$ and $t = t_1 * y * t_2$

and $a = c_1 * g(\forall) * y * c * c_2]\}$,

in coming to be, as the achievement of a prover's lived-work, the naturally accountable account of the work of its proving, is, as that achieved pairing of material proof and practice, the naturally accountable definition/proof of freefor(t, x, a).

Consider the following drawing taken from a book on Gestalt psychology:

From *Principles of Gestalt Psychology* by K. Koffka, copyright © 1935 by Harcourt Brace Jovanovich, Inc., renewed 1963 by Elizabeth Koffka. Reproduced by permission of the publisher.

The point of this drawing is that although, in some sense, it is a 'chaotic jumble of lines,'[132] the lines of that drawing provide enough detail so that they are seen as composing, and are organized as, the face of a plumpish, spectacled man. The analogy to notation's work is this: in the definition/proof of freefor(t, x, a), the notation that is used serves to articulate the work to which that definition/proof, as the naturally accountable definition/proof of freefor(t, x, a), is irremediably tied. The point is that, from within the lived-work of proving, the achievement of notation's work is not that the notation provides a correspondence between symbol and referent nor, for that matter, that it provides a correspondence between symbol and practice, but that — in the way in which the orderliness of the material detail of a proof and that orderliness's associated work are mutually developed and intrinsically complementary, and in the way in which that complementarity is itself a developing and produced feature of the work of proving — a proof's notation comes to articulate just those practices to which the naturally accountable proof is itself tied.

The material in this section can now be drawn together. First, if we understand by a schedule of proofs' notation the accountable orderliness of that proof's material detail, then that orderliness — as a prover's achievement in constructing or working through a schedule of proofs — articulates (in the sense of an intrinsic complementation) the work of producing the naturally accountable schedule of proofs. This being the case, the fact that the designation Sub(x, t, a) offers itself (again, as its achievement) as providing a correspondence between its 'accountable detail' and that detail's intended referent is itself a disengaged residue of notation's work — that work only being available at the mathematical work-site and hidden even there in and as that work's accountable accomplishment. And, last, if we now understand that a reference to the consistency of a schedule of proofs' notation is, in fact, a reference to that notation's exhibited and accountable orderliness, then a prover's work in providing such a consistent notation is the work of locally articulating just those practices which, paired with the schedule's material detail, make up the coherence of that schedule as a naturally accountable course of provings' work.

Before leaving the discussion of a schedule of proofs' notation, it is now possible to address one of the notational details of the schedule of proofs (as it is presented in this book and in Robbin's *Mathematical Logic* as well) that may have troubled the reader. As the reader will recall, the proof of the primitive recursiveness of the function $n \mapsto g(x_n)$ enumerating the Gödel numbers of the individual variables of P was previously given as

$$g(x_n) = g((x) * \prod_{i=0}^{n} p_i^7 * g(\ )),$$

where x_n is an abbreviation for $(\overbrace{x'' \ldots}^{n}\ ')$ and 7 is the Gödel number that has been assigned to the primitive symbol '. The troublesome

A SCHEDULE OF PROOFS

aspect of this proof is only alluded to by pointing out that this proof is the only place in our schedule where the actual Gödel number of a symbol appears; throughout the rest of the entire schedule the introduction of the particular numerical assignments of the proof-specific Gödel numbering has been avoided through the use of the function, designated by g, that maps the expressions (but not the primitive symbols or sequences of expressions) of our formal system P onto their Gödel numbers. What is troublesome about the use of the number 7 is this: in the way in which the rest of our schedule of proofs does not explicitly use the actual assignment of specific Gödel numbers, that schedule works out and exhibits the relevant *structure* of that numbering for a proof of Gödel's theorem.[133] Against this background, the use of the specific Gödel number of ′ in the demonstration of the primitive recursiveness of the function $n \mapsto g(x_n)$ offers the possibility that a proof of Gödel's theorem may be dependent on just such a particular assignment of Gödel numbers.

Now, given that the presence of the 7 in the displayed proof, and given as well that it is 'apparent'[134] to a prover that the schedule of proofs is not so dependent on this particular assignment of Gödel numbers, the question is thereby raised as to why the 7 still appears in the proof.

The answer to this question is found through an examination of the notational work that would be involved, and the examination of the notational apparatus that would have to be built, so that that 7 could be removed.

As the reader will recall, the symbol g was defined as, and has been used throughout our work in developing the schedule of proofs to denote, the mapping of the *expressions* of P onto their Gödel numbers. However, in the proof of the primitive recursiveness of $n \mapsto g(x_n)$

$$g(x_n) = g((x) * \prod_{i=0}^{n} p_i^7 * g(\ )),$$

the Gödel number of the primitive symbol ′ (i.e., 7) and not the Gödel number of the expression consisting of ′ alone (i.e., 2^7) is needed. Now, the reason that the function g cannot be used to give this number — the reason that g cannot be used to denote the Gödel numbering as it is defined on the entire collection of primitive symbols, expressions and sequences of expressions of P — is that a Gödel numbering must assign different numbers to ′ when ′ is considered as a primitive symbol, as an expression, and as a sequence of expressions of P. The detail of the symbolism g(′) fails to distinguish between these different ways of interpreting ′.

Given this circumstance, there seem to be two ways to avoid the presence of 7 in the proof of $n \mapsto g(x_n)$: One way would be to write the exponent of p_i as $(g(′))_i$ since $g(′) = 2^7$ (the Gödel number of the

147

expression $'$) and, then, $(g('))_1$ evaluates $g(')$ at the exponent of p_1 in $g(')$, $(g('))_1 = (2^7)_1 = 7$. A second method would be to use some notation, like 'corners'[135] ⌈ ⌉, to indicate the Gödel numbers of the primitive symbols of P. However, in that the proper introduction of that notation would have to occur before its immediate use in the schedule[136] — thereby providing for the independence of the distinction between ⌈ ⌉ and g from the locally contingent details of the work of proving Gödel's theorem — the introduction of a special designation for the Gödel numbers of primitive symbols in addition to such a designation for the Gödel numbers of expressions would necessitate, as well, an additional notational designation for the Gödel numbers of sequences of expressions. A prover might, for example,[137] use g, g^+, and g^{++}, respectively, for these three functions.

These two devices suffer the following difficulties: In the first case, the decipherment of $(g('))_1$ exhibits the artificiality of first raising the Gödel number of the symbol $'$ to an exponent and then determining what that exponent is. The use of $(g('))_1$ not only obfuscates the structure of the proof by making a prover decipher it, but it then exhibits its artificiality in that self-same decipherment. In that a specific Gödel number has already been introduced in defining that numbering, it is easier and less contrived to simply use that assigned number. In the second case, if the reader is offered the distinction between g, g^+, and g^{++} — and, thereby, the relevance of that distinction to the work of proving Gödel's theorem — that notation then requires a prover to do more work on reading and working through the schedule of proofs. Every occurrence of g in the present schedule would have to be changed to g^+, yet the only material motivation for introducing that distinction is the single occurrence of the 7 in the proof of the primitive recursiveness of $n \mapsto g(x_n)$.[138] In this way, the distinction between g, g^+, and g^{++} would only distract from the clarity of the material presentation in providing for more work than is required to find the material presentation of the schedule as a naturally accountable schedule of proofs.

In summary, then, we have found not only the material motives for using the 7 in the proof of the primitive recursiveness of $n \mapsto g(x_n)$, but, at the same time, the advantage, in that it reduces the burden of a schedule's notation, of simply defining one function g defined exclusively on the expressions of P.

8 A Structure of Proving

A
The Characterization Problem: The Problem of Specifying What Identifies a Proof of Gödel's Theorem as a Naturally Accountable Proof of Just That Theorem; The Texture of the Characterization Problem and the Constraints on Its Solution; The Characterization Problem as *the* Foundational Problem

Over the course of the preceding chapters we have come to rediscover, in increasing, technical detail, the natural accountability of a proof of Gödel's theorem as being irremediably tied to the local, lived-work of that proof's production. In doing so, we have therein come to rediscover a proof of Gödel's theorem *as* a naturally accountable proof. One major problem remains, however, for the analysis of a proof of Gödel's theorem as lived-work: what is it that identifies a proof of Gödel's theorem, over the course of its production, from within the developing course of its own lived-work, as a naturally accountable proof of just that theorem? What is it that is identifying of the local work of its proof, not as the work of proving, but as the work of proving Gödel's theorem?

In this chapter I will refer to the problem of specifying what identifies a particular proof of Gödel's theorem as a naturally accountable proof of just that theorem as the 'characterization problem.' By a 'solution' to that problem, I will refer here to a descriptive analysis of the work of producing a proof of Gödel's theorem that provides, in and as the inspectable details of the lived-work of that theorem's proof, technical access to that proof's characterization problem as a problem, for provers at the mathematical work-site, in the local production of social order. The aim of this chapter is to begin to develop such a solution, and, as a first step in that development — as well as a means of indicating the texture and richness of the characterization problem itself — the present section indicates some of the immediate constraints that can be placed on what a solution to the characterization problem could be. The analyses of the preceding chapters allow this to be done in a brief and summarizing fashion.

To begin, then, a first constraint on a solution to the characterization problem is this: the characterization problem cannot be solved by construing the natural accountability of a proof of Gödel's theorem as a property inherited from, or as a reflection of the apodictic character of, an ideal or idealized proof of Gödel's theorem. On one hand, there is nowhere available such an ideal, transcendental, self-evidential and self-evidentially complete proof to consult; on the other, it is the materially present, practically objective, naturally accountable proof of Gödel's theorem that provides the grounds for speaking of a proof of Gödel's theorem as being disengaged from the work of its production and review in the first place. Similarly, the characterization problem cannot be solved by positing the transcendental existence of the mathematical objects about which that proof is concerned and then using that claimed existence to provide for the self-evidential character of those object's properties. In that it is the naturally accountable proof that makes available and exhibits both itself and those mathematical objects as 'real,' propertied objects, a solution to the characterization problem begins not by taking that natural accountability for granted, but by examining the work that makes up that natural accountability as an achievement.

A second and closely related constraint to be placed on a solution to the characterization problem is that such a solution cannot make use of an idealized version of mathematical reasoning to, in that way, insure the transcendental properties of a naturally accountable proof of Gödel's theorem. The reduction of the reasoning of such a proof to, for example, instances of the use of *modus ponens*[1] not only trivializes the reasoning of the proof, but irremediably hides the distinctiveness of the discovered structure of reasoning that that proof makes available. The rendering of a naturally accountable proof as a derivation within a formal system — and, therein, the purported demonstration of the disengaged character of that proof's reasoning from its lived-work — cannot find the reasoning of that proof particularly; in all but the most elementary cases, the work of finding such a rendering is absurdly complicated;[2] and in the case of a proof of Gödel's theorem, when an (abbreviated) rendering is even attempted, that rendering is produced, and becomes intelligible, only by using a naturally accountable proof of Gödel's theorem as a guide to that rendering's interpretation, that rendering itself being tied to the local work practices of its own production to insure its accountability as a rendering of *the* proof of Gödel's theorem.[3]

By disallowing the characterization problem to be solved by reference to a disengaged version either of a proof of Gödel's theorem, of

A STRUCTURE OF PROVING

such a proof's reasoning, or of the objects that are considered in such a proof, the first two constraints serve to relativize the characterization problem as a problem arising from within, and whose solution is maintained through, a prover's local organization of a materially-specific course of work as this-particular proof of Gödel's theorem. In this way, then, the first two constraints point to a third one as well: that the practical objectivity and practical accountability of a proof of Gödel's theorem, as the proof of just that theorem, are temporally-developing features of the work of producing and reviewing such a proof.

One of the ways in which a solution to the characterization problem can be provided for is by using a comparison between different proofs of Gödel's theorem to argue that the variations between those proofs are inessential features of them and, therefore, that a particular proof of Gödel's theorem is, in fact, the material-specific realization of an objective, work-transcendent proof that has been fitted to the circumstantial particulars of its immediate production. The constraints that have already been introduced indicate the irrelevance of (the posited existence of) such a transcendental proof to the solution of the characterization problem – a solution must provide for the recognizability and identifiability of a proof of Gödel's theorem in and as the local, temporally-situated, temporally-developing, endogenously organized work of a material-specific proof's production and review. Nevertheless, the existence of variations between proofs of Gödel's theorem point to a phenomenon internal to the production of a particular proof, namely that provers will recognize the features of such a proof, over the course of that proof's development, as being peculiar to that particular proof; provers will see a developing proof as a particular way of proving Gödel's theorem. A mathematician watching a student work through a proof of the theorem will, at times, recognize the student's work as being faulted, plodding or disorganized, or will recognize that the material detail of the student's work is unmotivated by the circumstances of the proof given so far, but, at the same time, he will see how the student's work can be reorganized so as to provide an adequate proof of Gödel's theorem. That a solution to the characterization problem provides for the transcendental character of the work of proving Gödel's theorem from within the local work of that proof's production is a fourth constraint on such a solution.

A fifth constraint is that a solution must provide for the intelligibility of three somewhat mysterious features of the work of producing such a proof: the first is that certain details of a proof, later available to a reviewer as being obviously necessary for that proof, will be curiously missed by the prover when he was constructing the proof; second, and conversely, from within the work of producing such a proof a prover will find it necessary to introduce certain details into that proof that, in a later review, he will see as unnecessary and even unmotivated by the encompassing proof; and third, the reasoning of part of the argument of the proof — while apparent from within the work of its development — may, on later review, seem unsuggestive and even unintelligible, and the prover will have to rework the reasoning of his argument to find the cogency, or the faulted character, of the argument that he had previously given. In summary then, this fifth constraint, in requiring that a solution to the characterization problem provide for the intelligibility of these three features of mathematical work, can be seen to raise the question of what constitutes the relevant and identifying detail of a proof of Gödel's theorem from within the work of proving, or the work of reviewing, that proof.

A sixth constraint is that the natural accountability and coherence of a course of proving as, identifiably, a proving of Gödel's theorem cannot be made to depend on a prover's ability to articulate the work of that proof in any other way than the way in which that work is done as recognizably adequate mathematical practice. The analyzability of a proof of Gödel's theorem arises from within and is essentially tied to professional mathematical praxis; it is not tied to the reflective availability of that practice in terms other than those provided by the practice itself. Despite the fact that, on reflection, a given proof never explicates every feature of its reasoning, despite the fact that such reflection offers the possibility of an endless regress of further explications, nevertheless, a given proof, as its achievement, will be recognizably the practically adequate, practically objective, practically complete proof of Gödel's theorem; in a proof's achievement as an exhibited, naturally accountable organization of mathematical practice, that proof will therein, simultaneously, exhibit the pointlessness of disputing its adequacy as such a proof.[4]

Finally, a seventh constraint to be placed on a solution to the characterization problem is that, despite — or against the background — of the other constraints of this list, a given proof of Gödel's theorem is, nevertheless, an 'organizational' or 'social object';[5] it is recognizably a practically objective, practically adequate proof of that theorem; it is an analyzable object with demonstrable properties; it has a transcendental presence as such a proof of Gödel's theorem; while apparently dependent on the material details of its presentation, it exhibits, as its achievement, its independence from the particular details of that presentation; at a given stage in the construction of a proof, the work that has been done in proving the theorem can be contrasted with, and constrains, the work that still remains to be done; a proof of Gödel's theorem is seen to have an atemporal character over the course of its temporal development and construction; for a mathematician proving Gödel's theorem, and for mathematicians party to such a proving, the work of proving that theorem is the witnessible and recognizably real thing that is being done in and as the material details of that work;

A STRUCTURE OF PROVING

that this is so — that Gödel's theorem is the real thing that is being proved — is not available to a mathematician as an idiosyncratic perception of his own or another prover's work, but as something that is objectively so about that work, as something that anyone[6] can see; the work of proving Gödel's theorem is available to the mathematician, from within the course of that work, as a matter of practical assessment, comment and review; and, finally, the way in which the proving is done is, in fact, a moral way of proving, it is recognizably a proper way to prove, it is recognizably a proper way to prove Gödel's theorem, and the proving is done in such a way so as to be, recognizably, just such a proper way of proving.

This, then, is a list of constraints that can initially be placed on a solution to the characterization problem. That they are actual constraints is not a matter of principle or of proper argumentation; they are constraints in the way in which they serve to summarize some of the features of a-proof-of-Gödel's-theorem/the-lived-work-of-its-production that a proof of Gödel's theorem, over the course of this book, has been disclosed to be. It is this object to which the term 'a proof of Gödel's theorem' refers, and it is the problem of specifying what identifies such a proof, to its local production cohort, from within and as the lived-work of its production, as a naturally accountable proof of Gödel's theorem, that makes up the characterization problem for a proof of Gödel's theorem. In this way, then, the characterization problem is, in fact, a radical problem in the production of social order.

In the next section I will begin to work toward a 'solution' to the characterization problem. It is clear, however, that while I have used the term 'characterization problem' to refer exclusively to the characterization problem for a proof of Gödel's theorem, the characterization problem and the constraints placed on its solution have a much greater relevance to the study of mathematicians' work. That this is so can be gained, in the first instance, from some of the problems involved in teaching beginning mathematics students 'what a proof consists of' and, therein, 'how to prove': each mathematical problem, no matter how similar to others, is nevertheless a distinctive problem, and a student must learn, in and as the finding of a given problem's relevant detail, how to form up a course of work that stands as a proof of that problem's solution; for each problem, it is impossible to specify beforehand what would make up a naturally accountable proof of it; even when the similarity between the proofs of two propositions is almost identical, this does not insure that a student will find what, as practice, that similarity consists of for proving the second proposition after having seen a proof of the first one;[7] proofs are not and (generally) cannot be found by consulting specifications of the rules of proper logical inference to find an appropriate way of proceeding; and while

A STRUCTURE OF PROVING

the 'vernacular' use of rules of logic are relevant to the adequacy of a proof's reasoning, the insistence on characterizing a proof as a scheme of deductive inference is a pedagogic device for teaching what, in the end, is known and available to a prover as mathematical praxis. If we add to these brief observations just two more — that a discovered proof is a discovered naturally accountable proof and that professional mathematicians, in proving theorems, are engaged in teaching colleagues the-particular-way-of-proving/the-particular-way-of-reasoning that makes up a proof of the theorem in question — then we come to see that the characterization problem — as the problem of specifying what it is about a particular proof that identifies *that proof* as the naturally accountable proof of just the theorem that that proof proves — can, in fact, now be understood, not as *a* foundational problem, but as *the* problem of the foundations of mathematics.

B
Generalizing the Proof of Gödel's Theorem (As a Means of Gaining Technical Access to the Characterization Problem)

In the remainder of this chapter I want to develop the notion that for provers a-proof-of-Gödel's-theorem/the-lived-work-of-its-proving is, and is recognizable and identifiable as, an organization of proving's practices. I will speak of this organization of practices as the 'structure of proving' of a proof of Gödel's theorem, and later in this chapter I am going to propose that this 'structure of proving,' available only in and as the lived-work of that proof's production, provides a provisional[8] solution to the characterization problem. Before doing so, this section will first address a technical problem in the proof of Gödel's theorem as that proof has been presented here — namely, the fact that the adequacy of that proof (for proving the thing that it accountably proves) depends on the witnessible generalizability of that proof's reasoning. By elaborating on this problem and its solution, this section provides background material for the chapter's larger argument.

Let us begin, then, by considering the work that remains for our proof of Gödel's theorem to be a practically adequate proof of that theorem. As it was stated earlier, Gödel's first incompleteness theorem says (roughly)[9] that if the formal theory designated by P is consistent, then neither the particular, seeably/showably exhibitable sentence J nor that sentence's negation $\sim J$ are deducible in P. Since a formal theory is defined to be complete if and only if, for every sentence[10] S of that theory, either S or $\sim S$ is deducible in it, then, by definition, the proof that neither J nor $\sim J$ is deducible in P — or, in other words, that J is a 'formally undecidable' sentence of P — shows that P is not a complete theory.[11]

A STRUCTURE OF PROVING

By putting together the various parts of the proof of Gödel's theorem as they have been developed to this point, we have thus far shown that P is not a complete theory. But this statement of Gödel's theorem is certainly problematic: as the reader will recall, P was originally constructed as a 'model' of the theory of the natural numbers (that is, as a model of elementary arithmetic), and 'Gödel's theorem,' as it has just been stated, does not assert that there exists an undecidable proposition of number theory, but only that there exists an undecidable proposition of P; the possibility remains that only P, and not number theory, is incomplete or, stated in a fashion more compatible with the Hilbertian program for foundational studies, that there exist different formalizations of number theory that are both consistent and complete. In fact, an immediate candidate for such a formal theory is simply our original theory P with the undecidable sentence J now added to P as an additional axiom.[12,13] By construction, J would be deducible in this new theory, and the problem would thereby arise of showing either that there exists or that there does not exist an undecidable proposition of it. Thus, our proof of Gödel's theorem for P 'could be seen'[14] to leave open the question of the applicability of that theorem to other logistic systems that contain 'models' of number theory; thus, to leave open the consequentiality of that theorem for the program of the formal investigation of number theory; thus, to leave open the consequentiality of it for the (classical) investigation of the foundations of mathematics in general.

This, of course, is not the case: the proof of Gödel's theorem demonstrates what is sometimes referred to as the essential incompleteness of P — namely, that the incompleteness of P cannot be remedied by adding to P, for example, any class of axioms the Gödel numbers of which make up a primitive recursive set — and, in fact, that proof makes available the essential incompleteness of a much larger class of formal systems modelling elementary arithmetic than those built from the particular system P.

Our problem, then, in descriptively analyzing the natural accountability of a proof of Gödel's theorem, is that of finding how the generalizability of Gödel's theorem is tied to the lived-work of that theorem's proof. As a means of doing this, I am going to review the entire proof of Gödel's theorem: axioms for the system P will be introduced, a Gödel numbering will be specified, a schedule of proofs will be given, and the diagonalization/'proof' will be presented once again. The aim of so forming up the various 'parts' of a proof of Gödel's theorem into a coherent proof is to exhibit the essential connection between the generalizability of that proof and the availability of that proof, to a prover, as a 'structure of proving.'

To begin then, let me introduce, although in an abbreviated manner,[15] a specific set of axioms and rules of inference for the formal system

155

A STRUCTURE OF PROVING

now to be denoted as the system P. For the primitive symbols of P let us take $(,)$, $'$, x, $\supset$, $\sim$, $\forall$, $=$, 0, S, $+$, and $\cdot$, and let us adopt the following as axioms and axiom schemata where A, B, and C are wffs of P:[16]

1. $A \supset (B \supset A)$
2. $((A \supset (B \supset C)) \supset ((A \supset B) \supset (A \supset C))$
3. $(\sim (\sim A)) \supset A$
4. $\forall xA \supset S_t^x A|$, where x is an individual variable and t is a term which is free for x in A
5. $\forall x(A \supset B) \supset (A \supset \forall xB)$, where x is an individual variable having no free occurrences in A
6.[17] $(x_1 = x_2) \supset (x_1 = x_3 \supset x_2 = x_3)$
7. $(x_1 = x_2) \supset (S(x_1) = S(x_2))$
8. $\sim (S(x_1) = 0)$
9. $(S(x_1) = S(x_2)) \supset (x_1 = x_2)$
10. $x_1 + 0 = x_1 =$
11. $x_1 + S(x_2) = S(x_1 + x_2)$
12. $x_1 \cdot 0 = 0$
13. $x_1 \cdot S(x_2) = (x_1 \cdot x_2) + x_1$
14. $S_0^x A| \supset (\forall x(A \supset S_{S(x)}^x A|) \supset \forall xA)$.

The rules of inference of P will be *modus ponens*

from A and $A \supset B$, to infer B

and generalization,

if x is an individual variable, from A to infer $\forall xA$.

Now, let us introduce a specific Gödel numbering for the language of P. The Gödel numbers of the primitive symbols of P are defined as follows:

(... 3
) ... 5
' ... 7
x ... 9
$\supset$ ... 11
$\sim$ ... 13
$\forall$ ... 15
$=$ ... 17
0 ... 19
S ... 21
$+$ ... 23
$\cdot$ ... 25

If $\alpha = s_1 \ldots s_n$ is a sequence of primitive symbols of P (i.e., α is an 'expression' of P) and s_i is the Gödel number assigned to the symbol s_i, then the Gödel number assigned to α is $g(\alpha) = p_1^{s_1} \cdot \ldots \cdot p_n^{s_n}$,

A STRUCTURE OF PROVING

where p_i is the i-th prime number, $p_0 := 1$, and the symbol g is used exclusively for the Gödel numbers of the expressions of P. If $\Lambda = \alpha_1, \ldots, \alpha_n$ is a sequence of expressions of P, then the Gödel number assigned to Λ is $p_1^{g(\alpha_1)} \cdot \ldots \cdot p_n^{g(\alpha_n)}$.

Before giving a 'schedule of proofs' for P, let us recall the definition of the class of primitive recursive functions and that of the class of primitive recursive relations. First, if h is an n-place numerical function and $g_1, \ldots, g_n$ are m-place numerical functions, then the m-place function f defined by

$$f(x_1, \ldots, x_m) = h(g_1(x_1, \ldots, x_m), \ldots, g_n(x_1, \ldots, x_m))$$

is said to be obtained from $g_1, \ldots, g_n$ and h by *substitution*. If g is an m-place numerical function and h is an (m + 2)-numerical function, then the (m + 1)-place function defined by

$$f(x_1, \ldots, x_m, 0) = g(x_1, \ldots, x_m)$$
$$f(x_1, \ldots, x_m, S(y)) = h(x_1, \ldots, x_m, y, f(x_1, \ldots, x_m, y))$$

is said to be obtained from g and h by *primitive recursion*. And if a numerical function is either the zero function (denoted by) Z, the successor function S, or one of the projection fuctions $I_i^m, 0 < i, 0 < m$,

$$Z(x) = 0$$
$$S(x) = x + 1$$
$$I_i^m(x_1, \ldots, x_i, \ldots, x_m) = x_i$$

that function is said to be an *initial function*. Then the *primitive recursive functions* are defined inductively as only those functions obtained from the initial functions by a finite number of substitutions or primitive recursions; less precisely, they are the functions generated from the initial functions by the 'operations' of substitution and primitive recursion.

By definition, an m-place numerical relation R is a *primitive recursive relation* if and only if its characteristic function K_R,

$$K_R(x_1, \ldots, x_m) = \begin{cases} 1 \text{ if } (x_1, \ldots, x_m) \in R \\ 0 \text{ if } (x_1, \ldots, x_m) \in N^m - R \end{cases},$$

is a primitive recursive function.

A number of functions and relations will now be shown to be primitive recursive, among them functions and relations corresponding under the Gödel numbering to the syntax of P. The list begins with some propositions indicating some elementary primitive recursive functions and relations and some procedures for constructing primitive recursive functions and relations (Propositions 1-10); the proofs of these propositions will not be given.[18]

1 The constant functions $Z_n(x) = n$, $n = 0, 1, 2, \ldots$ are primitive recursive.

2 The functions obtained by permuting variables, identifying variables, adding dummy variables, and substituting constants for variables in primitive recursive functions are primitive recursive.

3 Addition, multiplication, exponentiation, and 'limited subtraction,' $\dotdiv$, are primitive recursive functions.
Let f be an $(m + 1)$-place numerical function and let $\sum_{k=0}^{y} f(x_1, \ldots, x_m, k)$ and $\prod_{k=0}^{y} f(x_1, \ldots, x_m, k)$ denote the functions

$(x_1, \ldots, x_m, y) \mapsto f(x_1, \ldots, x_m, 0) + \ldots + f(x_1, \ldots x_m, y)$

or

$(x_1, \ldots, x_m, y) \mapsto f(x_1, \ldots, x_m, 0) \cdot \ldots \cdot f(x_1, \ldots, x_m, y),$

respectively.

4 If f is a primitive recursive function, then so are $\sum_{k=0}^{y} f(x_1, \ldots, x_m, k)$ and $\prod_{k=0}^{y} f(x_1, \ldots x_m, k)$.

5 $=, \neq, <$, and $\leq$ are primitive recursive relations.

6 The logical operations of 'not,' 'and,' 'or,' 'implies,' and 'if and only if,' applied to primitive recursive relations, produce primitive recursive relations.

7 The relation obtained by substituting a primitive recursive function for a variable in a primitive recursive relation is primitive recursive.
Let R be an $(m + 1)$-place numerical relation and let $\forall z \leq y\ R(x_1, \ldots, x_m, z)$ and $\exists z \leq y\ R(x_1, \ldots, x_m, z)$ denote the $(m + 1)$-place relations that hold for $(x_1, \ldots, x_m, y)$ if and only if $R(x_1, \ldots, x_m, z)$ holds for all $z \leq y$ or for some $z \leq y$, respectively.

8 If R is a primitive recursive relation, then so are $\forall z \leq y\ R(x_1, \ldots, x_m, z)$ and $\exists z \leq y\ R(x_1, \ldots, x_m, z)$.

9 If $T_1, \ldots, T_v$ are pairwise disjoint m-place primitive recursive relations, and if $g_1, \ldots, g_v$ and h are m-place primitive recursive functions, then the function defined by the equation

$$f(x_1, \ldots, x_m) = \begin{cases} g_1(x_1, \ldots, x_m) \text{ if } T_1(x_1, \ldots, x_m) \\ g_2(x_1, \ldots, x_m) \text{ if } T_2(x_1, \ldots, x_m) \\ \quad \vdots \\ g_v(x_1, \ldots, x_m) \text{ if } T_v(x_1, \ldots, x_m) \\ h(x_1, \ldots, x_m) \text{ if } (x_1, \ldots, x_m) \notin T_i \text{ for all i}, 1 \leq i \leq v \end{cases}$$

A STRUCTURE OF PROVING

is primitive recursive.

If R is an (m + 1)-place numerical relation, let $\mu z \leq y\ R(x_1, \ldots, x_m, z)$ denote the function of $(x_1, \ldots, x_m, y)$ defined by the equation

$$\mu z \leq y\ R(x_1, \ldots, x_m, z) = \begin{cases} \text{the least } z \leq y \text{ such that} \\ \quad R(x_1, \ldots, x_m, z) \text{ if there is} \\ \quad \text{such a } z \\ 0 \text{ otherwise.} \end{cases}$$

10 If R is an (m + 1)-place primitive recursive relation, then $\mu z \leq y\ R(x_1, \ldots, x_m, z)$ is a primitive recursive function.

11 The 2-place relation x|y is primitive recursive.

Proof: $x|y \Leftrightarrow \exists n \leq y\ (y = n \cdot x)$.

12 Let prime(x) hold if and only if x is a prime number. Then prime(x) is a primitive recursive relation.

Proof: prime(x) $\Leftrightarrow x > 1$ and $\forall y \leq y\ (y|x$ implies $(y = 1$ or $y = x))$.

13 The function p_n giving, for each n, the n-th prime number is primitive recursive. ($p_0 := 1$.)

Proof: $p_0 = 1$

$$p_{n+1} = \mu x \leq (p_n)^n + 1\ \{\text{prime}(x)\ \text{ and }\ p_n < x\}.$$

To see that $(p_n)^n + 1$ is an upper bound on x, it is enough to note that $(p_1 \cdot \ldots \cdot p_n) + 1$ either is a prime number or is divisible by some prime greater than p_n, for it then follows that $p_{n+1} \leq (p_1 \cdot \ldots \cdot p_n) + 1 < (p_n)^n + 1$.

14 Define $(x)_n$ as the exponent of p_n in the prime factorization of x if $x > 1$ and $n > 0$ and as 0 otherwise. Then $(x)_n$ is a primitive recursive function of x and n.

Proof: $(x)_n = \begin{cases} 0 \text{ if } x = 0 \text{ or } i = 0 \\ \mu k \leq x\ (p_n^k | x\ \text{ and }\ p_n^{k+1} \nmid x) \text{ otherwise.} \end{cases}$

15 Let the function L(x) give the number n of the largest prime p_n in the prime factorization of x or give 0 if x is 0 or 1, i.e., L(x) defines the 'length' of x. L(x) is a primitive recursive function.

Proof: $L(x) = \mu n \leq x\ (p_n | x\ \text{ and }\ \forall k' \leq x\ (n < k \text{ implies } p_k \nmid x))$.

For every natural number y, $y = 0$ or $y = p_0^{(y)_0} \cdot p_1^{(y)_1} \cdot \ldots \cdot p_{L(y)}^{(y)_{L(y)}}$. Define x * y as the function mapping (x, y) to the value

$$x * y = x \cdot \prod_{i=0}^{L(y)} p_{L(x)+1}^{(y)_i}.$$

16 x * y is a primitive recursive function of x and y.

159

It follows from the definition that $x * (y * z) = (x * y) * z$ for all x, y, and z greater than 0. Thus, the finite 'product' of natural numbers $a_i > 0$, $i = 1, \ldots, n$, can be written unambiguously as $a_1 * \ldots * a_n$.

17 The function $n \mapsto g(x_n)$ giving, for each n, the Gödel number of the n-th individual variable, is primitive recursive.

Proof: $g(x_n) = g((x) * \prod_{i=0}^{n} p_i^7 * g( ))$.

18 Let var(x) be the relation that holds if and only if x is the Gödel number of a variable. var(x) is a primitive recursive relation.

Proof: var(x) $\Leftrightarrow \exists n \leq x \, (x = g(x_n))$.

Define a formation sequence of terms as a finite sequence of terms $t_1, \ldots, t_i, \ldots, t_m$ such that, for each i, $i = 1, \ldots, m$, one of the following conditions holds:

(i) t_i is 0
(ii) t_i is an individual variable
(iii) t_i is $S(t_j)$ for some $j < i$
(iv) t_i is $+(t_j t_k)$ for some $j < i, k < i$
(v) t_i is $\cdot(t_j t_k)$ for some $j < i, k < i$.

19 formterm(τ), holding if and only if τ is the Gödel number of a formation sequence of terms, is a primitive recursive relation.

Proof:

formterm(τ) $\Leftrightarrow \tau \neq 0$ and $\tau \neq 1$ and

$\forall i \leq L(\tau) \, \{[i = 0] \text{ or } [(\tau)_i = g(0)]$ or

$[\text{var}((\tau)_i)]$ or $\exists j < i \, [(\tau)_i = g(S( ) * (\tau)_j * g( ))]$

or $\exists j < i \, \exists k < i \, [(\tau)_i = g(+( ) * (\tau)_j * (\tau)_k * g( ))]$

or $\exists j < i \, \exists k < i \, [(\tau)_i = g(\cdot( ) * (\tau)_j * (\tau)_k * g( ))]\}$.

20 Let term(t) hold if and only if t is the Gödel number of a term of P. term(t) is a primitive recursive relation.

Proof: term(t) $\Leftrightarrow \exists \tau < \prod_{i=0}^{L(t)} p_i^t \, \{\text{formterm}(\tau) \text{ and } [(\tau)_{L(\tau)} = t]\}$.

To see that the bound on τ is adequate, let t be the Gödel number of a term $t = s_1 \ldots s_i \ldots s_m$, where the s_i are primitive symbols of P. By reviewing the definition of a formation sequence of terms, the reader will see that there exists at least one formation sequence for t of length less than or equal to m. If τ is the Gödel number of such a formation sequence, then $\tau \leq p_1^t \cdot \ldots \cdot p_m^t = \prod_{i=0}^{L(t)} p_i^t$.

A STRUCTURE OF PROVING

In analogy with the definition of a formation sequence of terms, let us call a finite sequence $A_1, \ldots, A_i, \ldots, A_m$ of wffs a formation sequence of wffs if and only if, for each A_i, $i = 1, \ldots, m$, one of the following conditions holds:

(i) A_i is of the form $(r = s)$ where r and s are terms
(ii) A_i is $(\sim A_j)$ for some $j < i$
(iii) A_i is $(A_j \supset A_k)$ for some $j < i$, $k < i$
(iv) A_i is $(\forall x A_j)$ for some individual variable x and for some $j < i$.

21 formwff(σ), which holds exactly when σ is the Gödel number of a formation sequence of wffs, is a primitive recursive relation.

Proof:

formwff(σ) $\Leftrightarrow$ $\sigma \neq 0$ and $\sigma \neq 1$ and $\forall i \leqslant L(\sigma)$ {[i = 0] or

$\exists t < \sigma \; \exists s < \sigma$ [term(t) and term(s) and

$(\sigma)_i = g((\;) * t * g(=) * s * g(\;))]$ or

$\exists j < i \;[(\sigma)_i = g((\sim) * (\sigma)_j * g(\;))]$ or

$\exists j < i \;\exists k < i \;[(\sigma)_i = g((\;) * (\sigma)_j * g(\supset) * (\sigma)_k * g(\;))]$ or

$\exists j < i \;\exists x < \sigma$ [var(x) and $(\sigma)_i = g((\forall) * x * (\sigma)_j * g(\;))$]

22 Let wff(a) hold if and only if a is the Gödel number of a wff of P. wff(a) is a primitive recursive relation.

Proof: wff(a) $\Leftrightarrow$ $\exists \sigma \leqslant \prod_{i=0}^{L(\sigma)} p_i^a$ [formterm(σ) and $(\sigma)_{L(\sigma)} = a$].

Define the relation occur(a_1, x, a_2, a) by the condition occur(a_1, x, a_2, a) $\Leftrightarrow$ wff(a) and var(x) and $a = a_1 * x * a_2$. occur(a_1, x, a_2, a) holds if and only if a is the Gödel number of a wff A, x is the Gödel number of a variable x that occurs in A, a_1 is the Gödel number of a part of the formula A to the left of an occurrence of x, and a_2 is the Gödel number of the part of the formula A to the right of that occurrence of x. Then

23 occur(a_1, x, a_2, a) is a primitive recursive relation.

24 Let bound(a_1, x, a_2, a) hold if and only if a is the Gödel number of a wff A, x is the Gödel number of a variable x which occurs in A, and the occurrence of the variable x in A indicated by occur(a_1, x, a_2, a) is a bound occurrence of that variable. Then bound(a_1, x, a_2, a) is a primitive recursive relation.

Proof:

bound(a_1, x, a_2, a) $\Leftrightarrow$ occur(a_1, x, a_2, a) and

$\exists c_1 \leqslant a \;\exists c_2 \leqslant a \;\exists c \leqslant a$ {wff(c) and

A STRUCTURE OF PROVING

$L(c_1) \leq L(a_1)$ and $L(c_2) \leq L(a_2)$ and
$a = c_1 * g(\forall) * x * c * c_2\}$.

Let the relation free(a_1, x, a_2, a) hold if and only if occur(a_1, x, a_2, a) and not-bound(a_1, x, a_2, a), i.e.

free(a_1, x, a_2, a) $\Leftrightarrow$ occur(a_1, x, a_2, a) and not-bound(a_1, x, a_2, a).

free(a_1, x, a_2, a) indicates a free occurrence of a variable x (corresponding to the number x) in a wff A (corresponding to the number a).

25 free(a_1, x, a_2, a) is a primitive recursive relation.

Define a function $S^1(x, t, a)$ as follows:

$$S^1(x, t, a) = \begin{cases} \mu b \leq t * a * t \; \{\exists a_1 \leq a \; \exists a_2 \leq a \; [\text{free}(a_1, x, a_2, a) \text{ and } \\ \qquad b = a_1 * t * a_2]\} \\ \qquad \text{if } \exists a_1 \leq a \; \exists a_2 \leq a \; [\text{free}(a_1, x, a_2, a)] \\ a \qquad \text{otherwise.} \end{cases}$$

Then if x is the Gödel number of a variable x, t is the Gödel number of a term t, and a is the Gödel number of a wff A, $S^1(x, t, a)$ gives the Gödel number of the wff resulting from the replacement of one free occurrence of x in A by t.

26 $S^1(x, t, a)$ is a primitive recursive function.

27 $S^n(x, t, a)$, iterating the operation of $S^1(x, t, a)$ n times, is a primitive recursive function of n, x, t, and a.

Proof: $S^0(x, t, a) = a$

$S^{n+1}(x, t, a) = S^1(x, t, S^n(x, t, a))$.

Define $M(x, t, a) = S^{L(a)}(x, g(x_{L(a)+L(t)}), a)$. $M(x, t, a)$ replaces all the free occurrences of x in A by a variable $x_{L(a)+L(t)}$ that does not occur either in t or in A.

28 $M(x, t, a)$ is a primitive recursive function.

Define the function Sub(x, t, a) by the equation

$$\text{Sub}(x, t, a) = \begin{cases} g(S_t^x A|) & \text{if } x = g(x), t = g(t), a = g(A), \\ & x \text{ an individual variable, } t \text{ a} \\ & \text{term, and } A \text{ a wff.} \\ a & \text{otherwise.} \end{cases}$$

29 Sub(x, t, a) is a primitive recursive function.

Proof: $\text{Sub}(x, t, a) = S^{L(a)}(g(x_{L(x)+L(a)}), t, M(x, t, a))$.

30 Define a function sub(x, n, a) by the equation

A STRUCTURE OF PROVING

$$\text{sub}(x, n, a) = \begin{cases} g(S^x_{k(n)} \, A \, |) & \text{if } x = g(x), a = g(A), x \text{ an} \\ & \text{individual variable and } A \text{ a wff,} \\ & \text{where } k(n) = \overbrace{S(\ldots S(0)}^{n} \ldots) \\ a & \text{otherwise} \end{cases}$$

sub(x, n, a) is a primitive recursive function.

Proof: First, g(k(n)) is a primitive recursive function of n by primitive recursion, since

$g(k(0)) = g(0)$

$g(k(S(n))) = g(S(\,) * g(k(n)) * g(\,))$.

Then, sub(x, n, a) = Sub(x, g(k(n)), a).

31 freefor(t, x, a) is a primitive recursive relation where freefor(t, x, a) holds if and only if t, x, and a are the Gödel numbers of a term t, a variable x, and a wff A, respectively, and if t is free for x in A.

Proof:

freefor (t, x, a) $\Leftrightarrow$ term(t) and var(x) and wff(a) and

$\exists a_1 \leqslant a \; \exists a_2 \leqslant a \;\; \{\text{free}(a_1, x, a_2, a) \text{ implies}$

not-$\exists c_1 \leqslant a \; \exists c_2 \leqslant a \; \exists c \leqslant a \; \exists y \leqslant a \; \exists t_1 \leqslant t$

$\exists t_2 \leqslant t \; [\text{wff}(c) \text{ and var}(y) \text{ and } L(c_1) \leqslant L(a_1)$

and $L(c_2) \leqslant L(a_2)$ and $t = t_1 * y * t_2$ and

$a_1 = c_1 * g(\forall) * y * c * c_2]\}$.

32 notfree(x, a), which holds if and only if x is the Gödel number of a variable x, a is the Gödel number of a wff A, and x has no free occurrence in A, is a primitive relation.

Proof: notfree(x, a) $\Leftrightarrow$ not-$\exists a_1 \leqslant a \; \exists a_2 \leqslant a \; [\text{free}(a_1, x, a_2, a)]$.

33.1 $\text{axiom}_1(w)$, which holds if and only if w is the Gödel number of an instance of the axiom scheme $(A \supset (B \supset A))$, is a primitive recursive relation.

Proof: $\text{axiom}_1(w) \Leftrightarrow \exists a \leqslant w \; \exists b \leqslant w \; \{\text{wff}(a) \text{ and wff}(b)$

and $w = g((\,) * a * g(\supset(\,) * b * g(\supset) * a * g(\,)))\}$.

.
.
.

33.4 $\text{axiom}_4(w)$, which holds if and only if w is the Gödel number of an instance of the axiom scheme $((\forall xA) \supset S_t^x A|)$, is a primitive recursive relation.

Proof: $\text{axiom}_4(w) \Leftrightarrow \exists a \leqslant w \; \exists x \leqslant w \; \exists t \leqslant w$ {wff(a) and

var(x) and term(t) and freefor(t, x, a) and

$w = g(((\forall) * x * a * g(\;) \supset) * \text{Sub}(x, t, a) * g(\;))\}$.

.
.
.

33.6 $\text{axiom}_6(w)$, which holds if and only if w is the Gödel number of axiom 6, is a primitive recursive relation.

Proof: $\text{axiom}_6(w) \Leftrightarrow w = g(((x_1 = x_2) \supset ((x_1 = x_3) \supset (x_2 = x_3))))$.

33.7 $\text{axiom}_7(w)$, which holds if and only if w is the Gödel number of axiom 7, is a primitive recursive relation.

Proof: $\text{axiom}_7(w) \Leftrightarrow w = g(((x_1 = x_2) \supset (S(x_1) = S(x_2))))$.

.
.

33.14 $\text{axiom}_{14}(w)$, which holds if and only if w is the Gödel number of an instance of the axiom scheme $(S_0^x A| \supset ((\forall x(A \supset S_{S(x)}^x A|)) \supset (\forall xA)))$, is a primitive recursive relation.

Proof: $\text{axiom}_{14}(w) \Leftrightarrow$

$\exists a \leqslant w \; \exists b \leqslant w \; \exists c \leqslant w \; \exists x \leqslant w$ {wff(a) and var(x) and

$b = \text{Sub}(x, g(0), a)$ and $c = \text{Sub}(x, g(S(x)), a)$ and

$w = g((\;(\;) * b * g(((\forall) * x * g((\;) * a * g(\supset) * c * g(\;)) \supset (\forall) *$
$x * a * g(\;))))\}$.

33 axiom(w), which holds if and only if w is the Gödel number of an axiom or of an instance of an axiom scheme, is a primitive recursive relation.

Proof: $\text{axiom}(w) \Leftrightarrow \text{axiom}_1(w)$ or . . . or $\text{axiom}_{14}(w)$.

34 Define $\text{deduct}(\lambda)$ as holding if and only if λ is the Gödel number of a deduction. $\text{deduct}(\lambda)$ is a primitive recursive relation.

Proof: $\text{deduct}(\lambda) \Leftrightarrow \lambda \neq 0$ and $\lambda \neq 1$ and

$\forall i \leqslant L(\lambda)$ {$i = 0$ or $\text{axiom}(\lambda)_i$) or

$\exists j \leqslant i \; \exists k \leqslant i \; [(\lambda)_k = g((\;) * (\lambda)_j * g(\supset) * (\lambda)_i * g(\;))]$

or $\exists j < i \; \exists x < \lambda \; [\text{var}(x) \text{ and } (\lambda)_i = g((\forall) * x * (\lambda)_j * g(\;))]\}$.

A STRUCTURE OF PROVING

The last two conditions correspond to the use of *modus ponens* and generalization.

35[19] The relation ded(λ, a), which holds if and only if λ is the Gödel number of a deduction of the wff with Gödel number a, is a primitive recursive relation.

Proof: ded(λ, a) $\Leftrightarrow$ deduct(λ) and $(\lambda)_{L(\lambda)} = a$.

The reader will recall that an m-place numerical relation W is *numeralwise expressible* in P if and only if there exists a wff $W(x_1, \ldots, x_m)$ of P with m free variables such that

(i) if $(a_1, \ldots, a_m) \in W$, then $\vdash_P W(k(a_1), \ldots, k(a_m))$
(ii) if $(a_1, \ldots, a_m) \notin W$, then $\vdash_P \sim W(k(a_1), \ldots, k(a_m))$

where $k(a) = \overbrace{S(\ldots S(0) \ldots)}^{a}$ is the numeral of P corresponding to the number a. We then have the following

Theorem.[20] Every primitive recursive relation is numeralwise expressible in P.

Finally, let us define P to be ω-consistent if there is no wff $F(x)$ of P such that both $\vdash_P \exists x F(x)$ and $\vdash_P \sim F(k(0)), \ldots, \vdash_P \sim F(k(n)), \ldots$ for all n.

The proof of Gödel's first incompleteness theorem for P is brought to a close with the following construction of the sentence J and the proof that that sentence is formally undecidable in P.

First, let $\varphi(u) = \text{sub}(g(x_2), u, u)$. $\varphi(u)$ is the primitive recursive function such that, if u is the Gödel number of a wff U, then $\varphi(u)$ gives the Gödel number of the wff that results when the numeral $k(u)$ representing u in P is substituted for x_2 in U. Next, define the primitive recursive relation $G(\lambda, u)$ by the condition that

$G(\lambda, u) \Leftrightarrow \text{ded}(\lambda, \varphi(u))$

$\Leftrightarrow \text{ded}(\lambda, \text{sub}(g(x_2), u, u))$.

Roughly, $G(\lambda, u)$ holds if λ is the Gödel number of a deduction of the wff that results when $k(u)$ is substituted for x_2 in the wff with Gödel number u. Since $G(\lambda, u)$ is primitive recursive, there is a wff that numeralwise expresses G in P. Let $G(x_1, x_2)$ be such a wff.

Define I as $\exists x_1 G(x_1, x_2)$ and let $g(I) = i$.

Define J as $\sim \exists x_1 G(x_1, k(i))$ and let $g(J) = j$.

Now, observe that $j = \varphi(i)$, that is, tht j is the Gödel number of the wff that results when $k(i)$ is substituted for x_2 in the wff with Gödel number i.

J can be interpreted as 'saying' that there does not exist a number n such that n is the Gödel number of a deduction of the formula that results when the numeral k(i) is substituted for x_2 in the wff with Gödel number i, or, in other words, J 'says' that J is not deducible in, and hence is not a theorem of, P.

The 'proof' of Gödel's theorem can now be given.

Gödel's First Incompleteness Theorem (for P). (1) If P is consistent, J is not deducible in P. (2) If P is ω-consistent, $\sim J$ is not deducible in P.

Proof: (1) If J is deducible in P — that is, if

(*) $$\vdash_P \sim \exists x_1 G(x_1, k(i)),$$

— then, for some r, ded(r, j). But since

ded(r, j) $\Leftrightarrow$ ded(r, φ(i)) $\Leftrightarrow$ G(r, i),

the numeralwise expressibility of G gives

$\vdash_P G(k(r), k(i))$,

from which it then follows that

(**) $$\vdash_P \exists x_1 G(x_1, k(i)).$$

(*) and (**) cannot both hold if P is consistent. Thus, if P is consistent, J is not deducible in P.

(2) Suppose that $\sim J$ is deducible in P, that is

(***) $$\vdash_P \exists x_1 G(x_1, k(i)).$$

Since P is assumed to be ω-consistent, it follows that P is consistent as well; hence J cannot be a theorem of P. But if J is not a theorem, then, by definition of ded(r, j),

G(n, i) $\Leftrightarrow$ ded(n, φ(i)) $\Leftrightarrow$ ded(n, j)

cannot hold for any n $\in$ N. Using the numeralwise expressibility of G, one obtains

$$\vdash_P \sim G(k(1), k(i))$$

.
.
.

(****) $$\vdash_P \sim G(k(n), k(i))$$

.
.
.

for all n $\in$ N. Together, (***) and (****) contradict the assumption of ω-consistency. Thus, if P is ω-consistent, $\sim J$ is not deducible in P.

This, then, brings to a close the proof of Gödel's theorem for the system P. Once again, it says, roughly, that if P is a consistent system, then there exists a sentence J of P such that J is formally undecidable

A STRUCTURE OF PROVING

in P — i.e., that there exists a sentence J of P such that neither J nor $\sim J$ are deducible in P. In other words, it says that if P is consistent, then it is incomplete. Moreover, in that the proof of the numeralwise expressibility of primitive recursive relations in P gives an inductive procedure for constructing formulas that numeralwise express those relations,[21] it is possible, in principle, to construct and exhibit the actual sentence J, although in practice such a construction is infeasible if not impossible.

Now, let us return to our original question concerning the generality of this proof.

As I reminded the reader earlier, P was initially constructed as a formal 'model'[22] of number theory. The central problem that Gödel's theorem is intended to address is not whether or not P is complete, but whether or not ordinary number theory itself is complete, or, more precisely, whether or not there is *any* formal 'model' $\mathscr{P}$ of number theory such that that 'model' can be shown to be complete.

A first step in addressing this problem is taken by realizing that, since the axioms and rules of inference of P 'represent' true propositions of number theory and general, or epitomizing characterizations of, methods of ordinary mathematical inference, then a first-order formal system constructed as a 'model' of number theory would certainly seem to have to include, among its theorems, all the theorems of P. By this reasoning, then, a complete 'model' of number theory would have to contain — or, in more conventional terminology, would have to be an 'extension' of — our original system P.[23]

Let us consider one such extension of P.

First, note that since the sentence J is not deducible in P, and since J can be intuitively interpreted as 'saying' about itself that it itself is not deducible in P, J is, intuitively, a true sentence of number theory. Let us add J as a new axiom (axiom 15) to the axioms of P thereby obtaining an extension P' of P. In that J is an axiom of this new system, J is certainly deducible in it, and, thus, if P is consistent, P' will properly contain the theorems of P. In other words, we have 'enlarged' our 'model' of number theory, and the question that now confronts us is whether or not this enlarged system P' is a complete one.

In order to show that P' like P, is an incomplete system, we can reason as follows: In that J is a constructable formula of P, we can, in principle, calculate the Gödel number $g(J)$ of J and insert the proposition

33.15 $axiom_{15}(w)$, which holds if and only if w is the Gödel number of J, is a primitive recursive relation.

into our already constructed schedule of proofs. Modifying Proposition 33 accordingly by adding a new condition to the definition of axiom(w) corresponding to $axiom_{15}(w)$, we next obtain, in a new Proposition 34, the primitive recursive relation $\text{ded}'(\lambda, a)$ which holds exactly when λ is the Gödel number of a deduction of the wff of P' with Gödel number a. Following the procedure in the proof of Gödel's theorem for P, we can then construct a primitive recursive relation $G'(\lambda, a)$ and use the numeralwise expressibility of primitive recursive relations to obtain a new relation $G'(x_1, x_2)$, corresponding to the previously constructed $G(x_1, x_2)$ of P, numeralwise expressing G' in P'. A new sentence J' can then be constructed in the diagonalization argument, and, with only the notational replacement of P' for P, J' for J, ded' for ded, G' for G, and G' for G, the 'proof' of Gödel's theorem given above for P then applies to the new system P'. Hence, we conclude that if P' is consistent, then it is incomplete.

The pattern of the preceding argument should be clear. If, for example, we next add J' to P' as an axiom, we obtain a new system P''. Following the same procedure as above, we can construct a new primitive recursive relation $\text{ded}''(\lambda, a)$ and can use this relation to construct a sentence J'' which is formally undecidable in P''. More generally, consider a new set of axioms for a first-order 'model' P^+ where P^+ is an extension of P — that is, where all the theorems of P are theorems of P^+ as well.[24] Then as long as it is possible to 'effectively determine' the deductions of P^+ through the use of the Gödel numbering — that is, as long as $\text{ded}^+(\lambda, a)$ is a primitive recursive relation — we can, following the same procedure given for P, construct an undecidable sentence J^+ of P^+. In this way, then, we come 'to see'[25] that the system P is not only incomplete, but *essentially incomplete* in the sense that the incompleteness of P cannot be remedied by the addition of a 'reasonable' (i.e., primitive recursive) set of axioms.

We can, in fact, go further — we can consider a system $\mathscr{P}$ that 'extends' P not only by adding additional axioms, but by extending the language of P through the addition of new constant, function, and relation symbols. For example, we might consider the system obtained from P by adding to it the 2-place relation symbol $<$, the 2-place function symbol E, and the axioms[26]

(i) $\forall x \, \forall y \, [x < S(y) \equiv (x < y \text{ or } x = y)]$
(ii) $\forall x \, (x < 0)$
(iii) $\forall x \, \forall y \, (x < y \text{ or } x = y \text{ or } y < x)$
(iv) $\forall x \, [E(x, 0) = S(0)]$
(v) $\forall x \, \forall y \, [E(x, S(y)) = E(x, y) \cdot x]$.

(The intended interpretation of $<$ and E in N is, of course, the inequality relation and the exponentiation function, respectively.) In a manner similar to the modification of the schedule of proofs that we made in adding J to P as an axiom, we can modify the schedule so as to accommodate the symbols $<$ and E and the axioms for them. In fact, in section 3(d) of the last chapter[27] the modifications in our schedule that are necessary for adding a countably infinite number of function symbols f_i^m, $1 \leq m$, $1 \leq i$, were already considered. In this way then, we have seen some of the variations that can be made in the logical system P that do not affect the 'structure' of the proof of Gödel's theorem, and we can further appreciate the *essential incompleteness* of P in the following manner: if $\mathscr{P}$ is 'any' consistent, formal system that extends P — where the precise meaning of 'any' is now to be understood as itself a discoverable and accountable feature of the proof of Gödel's theorem — then $\mathscr{P}$ is incomplete.

C
A Structure of Proving: The Availability to a Prover of the Proof of Gödel's Theorem as a Structure of Practices; The Proof as the Pair The-Proof/The-Practices-of-Proving-to-Which-That-Proof-is-Irremediably-Tied

The aim of this section is to briefly summarize the discussion of the lived-work of proving Gödel's theorem and of the proof of that theorem as lived work.

In the preceding section, I reviewed the proof of Gödel's theorem for the formal system P, and then indicated how that proof can be generalized to any 'reasonable' extension of P. Following the publication of Gödel's original paper in 1931, Gödel's work in proving his incompleteness theorem for a simple theory of types was (in a certain sense) sharpened by Rosser[28] so as to avoid the necessity of assuming ω-consistency; Gödel's notion of 'rekursiv' functions, as found in the 1931 paper, was reformulated (in part, by Gödel himself)[29] as defining the subclass of primitive recursive functions of a larger class of functions, this larger class then being designated as the recursive functions (proper); the notions of 'numeralwise expressibility,' 'representability,' 'definability,' and 'decidability' were developed and applied; Church[30] showed that the set of theorems of formal number theory — i.e., the set of Gödel numbers of the theorems of formal number theory — is undecidable — i.e., not recursive; Tarski[31] demonstrated that the syntax of a formal system like P is not adequate to express its own semantics — i.e., that the set of Gödel numbers of the wffs of formal number theory that are 'true' in N is not 'arithmetical'; the question was raised as to the generalizability of Gödel's theorem — i.e., the question of which formal systems have results comparable to the incompleteness theorem; new topics were developed such as the notion of degrees of unsolvability; Gentzen[32] was able to prove that a different formulation of formal number theory was consistent, opening the possibility of the reinterpretation of Hilbert's notion of 'finitary' methods for metamathematical investigations, etc., etc.

This brief indication of the tremendous consequentiality of Gödel's work for, and of the pervasive influence of his work on, classical studies of mathematical practice might invite us into the reflection that the 'meaning' of Gödel's theorem and of its proof have an historical and historically unfolding character, that Gödel's work was motivated by, and set within, a specific historical context, that the ways in which we look at and interpet that theorem today is influenced by the various ways in which that theorem is embedded in its received histories, and the like. But it is against this envisioned historical flux that the proof of Gödel's theorem, then as now, stands as a recognized and recognizable achievement as the naturally accountable proof of the theorem that it is seen, from within the lived-work of its proof, to witnessibly and demonstrably prove. As I have tried to show over the course of this book, that achievement cannot be attributed to a work-transcendent state of mathematical objects, but, instead, the origins and substance of it must be found within the lived-work of proving Gödel's theorem itself. This problem was formulated in Section A of this chapter as the problem of specifying, as lived-work, what identifies a proof of Gödel's theorem as the proof of just that theorem — what, as practice, is identifying of that proof as its own demonstrably exhibited achievement. If we understand by a 'solution' to this problem a formulation of it that provides further descriptive and technical access to the problem itself as a problem, for provers, in the local production of social order, then, in the remainder of this section, I will try to suggest such a possible solution.

Let us return to the fact that the proof of Gödel's theorem for the system P can be 'generalized' to any 'reasonable' extension $\mathscr{P}$ of P. The first point that needs to be made is that this sense of generalization is not the same as the generalizations, sharpenings, modifications and reformulations of that theorem and its proof that followed the publication of Gödel's original paper. As I tried to indicate in the preceding section, the importance of Gödel's theorem (for P) is not just that the system P is incomplete, but that P is essentially incomplete, and the generalizability of the proof of Gödel's theorem for P to extensions of that system is itself an integral and necessary part of the proof of Gödel's theorem.

Now the reader will recall from Section B that the generalization of Gödel's theorem to extensions of P was obtained by first proving Gödel's theorem for P and then indicating how that proof provided for both the changes that were needed, and the modifications in the proof that were needed to incorporate those changes, for such a generalization. In this way, then, in that the generalizability of the proof is made available to the prover by his finding/being-directed-to-find the extractable, accountable structure of the proof's original argument, the proof of Gödel's theorem has an explicitly self-referential character: the proof of Gödel's theorem closes by pointing to the accountable

A STRUCTURE OF PROVING

structure of that proof such that that proof is applicable to 'any' formalization of number theory having a 'reasonable' set of axioms.

The important point that needs to be made is not that our proof of Gödel's theorem has such an explicitly self-referential character,[33] but that that proof-specific self-reference points to the availability, to a prover, of the proof of Gödel's theorem as a structure of proving — that is, as an organization of the practices of proving for proving-again Gödel's theorem. Summarily, that structure of proving can be construed in terms of the introduction of, and the intentionality of the introduction of, a Gödel numbering; the fact that that numbering allows the syntax of a formal system to be 'arithmetized'; the introduction of primitive recursive functions and relations and the construction of the arithmetized syntax as consisting of such functions and relations; the fact that, by showing that primitive recursive relations are numeralwise expressible in the formal system, the arithmetized syntax can be expressed in the syntax itself and that, in consequence, the diagonalization argument/'proof' can be given. The important point, however, is not this list of topics, but that this list is, in fact, a reference to an exhibited, orderly course of proving; it is a reference to the actual lived-work of proving-again Gödel's theorem. These topics and their arrangement provide an indication of how to prove Gödel's theorem — of a way of working that proves Gödel's theorem — and it is the availability to a prover of the proof of Gödel's theorem as such a structuring of the practices of proving that make it pointless to dispute that proof's witnessed achievement.[34]

As an initial summary, then, we can say that a proof of Gödel's theorem, as a structure of proving's practices, provides a course of practical reasoning and action — namely, the course of reasoning and action that is the identifying, naturally accountable, lived-work of proving-again Gödel's theorem. In this way, the proof of Gödel's theorem is a pedagogic object — it teaches provers how to prove Gödel's theorem, and it does this by providing, metaphorically, in the material proof, a template of that course of action. More descriptively, however, we must ask how that organization of practices is achieved in and as the detailed, lived-work of proving Gödel's theorem. What can now be proposed is this: a proof of Gödel's theorem *is* itself an intrinsically paired object[35] the-material-proof/the-practices-of-proving-to-which-that-proof-is-irremediably-tied; a proof of Gödel's theorem is produced as such a paired object; it is recovered as such a paired object; and it is in the realization of a proof of Gödel's theorem as such a paired object — and, therein, in the way in which the material proof is constructed as, and comes to stand as, an account of the work of its production, thereby articulating the proof-pair as such a pairing of material detail and practice — that a proof of Gödel's theorem is simultaneously realized as a structure of practices for proving, identifiably, just that theorem.

171

Part III Conclusion

9 Summary and Directions for Further Study

A
Classical Studies of Mathematical Practice: A Review of the Book's Argument[1]

In the introduction of this book, I introduced the notion of classical studies of mathematical practice — that is, of studies of the rigor, or natural accountability, of mathematicians' work for which the essentially local character of that accountability is irremediably hidden in and as the local work practices that make up the natural accountability of those studies themselves. I then proposed that 'conventional' studies of the foundations of mathematics are, in fact, such classical studies. In order to develop this proposal, it was necessary to gain real-world access to the rigor of a mathematical argument as a local phenomenon, and the body of this book was an attempt to do this through an extended descriptive analysis of the lived-work of proving one of Gödel's incompleteness theorems.

The book's argument was developed, roughly, as follows: among studies of the foundations of mathematics, Kurt Gödel's two incompleteness theorems are among the most celebrated. The first of these theorems 'says' that any consistent formal system strong enough to 'represent' elementary arithmetic is incomplete — that is, roughly, that there are true propositions of arithmetic that cannot be proved in such a formal system. Gödel's second incompleteness theorem 'says' that the consistency of such a formal representation of arithmetic cannot be demonstrated within that system itself or, even more loosely, that the consistency of mathematics cannot be proved mathematically. In attempting to give a precise description of these results in an early chapter, we found that the accuracy of such a description is tied to the material detail and lived-work of actually proving those theorems, and, thus, even more importantly, we found that the interpretability of those theorems rested on the prior availability of their proofs as naturally accountable mathematical arguments. The specification of what, as praxis, makes up the naturally accountable rigor of the proofs of Gödel's theorems is necessarily prior, as a foundational question, to the interpretation of those theorems in terms of mathematical practice.

The body of this book was devoted to an extended analysis of the lived-work of proving Gödel's first incompleteness theorem. Through that analysis I argued in an evidential manner that the rigor of a proof of Gödel's theorem consists of its local, lived-work — that a proof of Gödel's theorem was produced in such a way so as to be the recognizably and analyzably adequate proof of just that theorem. Although this argument was made through the particular analysis of the work of proving Gödel's theorem, what was uncovered as the work of *that* proof resonates throughout mathematical practice: by attending to the work of proving Gödel's theorem particularly, we gained material and technical access to the work of proving. The generality that was uncovered was not one of findings, but of the discovered existence of a phenomenon — that of the naturally accountable proof. In this way, then, the analysis of the proof of Gödel's theorem pointed to the discoverable origins of mathematical rigor in and as the lived-work of doing naturally accountable ordinary mathematics and to mathematical rigor as itself a problem, for provers, in the local production of social order.

To this point in our analysis we have treated the proof of Gödel's theorem as a proof of ordinary mathematics independently of its 'metamathematical' interpretation. Gödel's theorem is, of course, one of the great studies of the foundations of mathematics, and it is at this point that the book's argument turns back on itself: in so far as conventional foundational studies — like Gödel's theorem and its proof — are mathematical, they themselves are irremediably tied to the practices of doing recognizably adequate, rigorous mathematics; as does ordinary mathematical practice, conventional studies of mathematical foundations hide the origins of mathematic rigor in and as the rigor of their own, local, lived-work. Thus, the book returns to its opening proposal, that 'conventional' studies of the foundations of mathematics are properly characterized and investigated, as praxis, as classical studies of mathematicians' work.

In addition to the proposal of the existence of classical studies of mathematicians' work, the argument of this book has a further consequence for the study of the relationship between mathematical practice and the classical study of that practice. We would like to know if there is something 'natural' in the pairing of classical studies and mathematical practice; we would like to know if the origins of classical studies lie within mathematical practice itself. We have seen in this book that the rigor of a mathematical proof not only lies within, but is hidden within, that proof's lived-work, that a proof consists of the pair the-material-proof / the-practices-of-proving-to-which-that-proof-is-irremediably-tied, and that a proof is cultivated so as to realize the material proof as a disengaged version, or account, of that proof's lived-work. In this way, a mathematical proof is itself a classical study of its own practices. Conjecturally, then, a mathematical proof is itself a classical study of practical action; it is a classical study of the work of mathematical theory proving, and mathematics, as a discipline, is a classical science of practical action and practical reasoning.[2]

B
Prospectus: Mathematicians' Work as Structure Building

In this book I have addressed the problem of the nature and constitution of mathematical rigor — often spoken of as the problem of the foundations of mathematics — as a problem, for provers, in the local production of social order, and I have attempted to show that so formulated the foundations of mathematics can be investigated in and as the inspectable details of mathematical practice. This book is the beginning and not the end of this project. What I hope to have demonstrated, however, besides the feasibility of this project, is that one of the tasks that does not remain is that of building philosophical, historical or sociological elaborations of this way of posing foundational questions. The material in this book provides, I hope, a starting point for the investigation, as a real-wordly researchable matter, of what makes up the natural accountability of mathematicians' work; it provides a starting point for the discovery of what mathematical proofs identifiably consist of as lived-work, and it provides a starting point for the further examination of the proposal that the rigor of a proof is irremediably tied to the work of its local production. Moreover, in that the discovery of a mathematical proof is the discovery of a naturally accountable way of proving, this book provides an initial means of animating the question of the nature of mathematical discovery, not by romanticizing or psychologizing the mathematician, but by providing access to a mathematical discovery as lived-work.

In bringing this book to a close, I want to suggest a topic for further exploration which, because of the singular concentration in this book on the descriptive analysis of the lived-work of proving Gödel's theorem, could not be developed in the book itself. The idea is this: for mathematicians, the heart of their profession is not theorem proving *per se*; that it is is itself a classical version of their work. What I wish to propose is that the sustaining life of professional mathematics lies in what might be called 'mathematical structure building' — that is, the envisionment and construction of structures of theorems and proofs that have their

SUMMARY AND DIRECTIONS

motivating origins in, and are directed to the development and reformulation of, a current state of mathematical practice. What at this time I can only suggest is that the naturally accountable work of theorem proving provides essential and unique access to mathematical structure building; without knowing, as praxis, the work of theorem proving, the discovery, construction, recognition and consequentiality for mathematical practice of such mathematical structures is impossible. In closing, then, the questions that I wish to raise are whether or not it is possible to find mathematical structure building in and as the lived-work of doing professional mathematics; whether or not, by using the now available material on the lived-work of naturally accountable proving, it is possible to discover a natural technology in mathematicians' situated inquiries into mathematical structures; and whether or not, if such a natural technology exists, that technology is integrally tied to creativity in the work of mathematical discovery.

Appendix

The Use of Ethnomethodological Investigations of Mathematicians' Work for Reformulating the Problem of the Relationship between Mathematics and Theoretical Physics as a Real-World Researchable Problem in the Production of Social Order[1]

Over the course of the past several years, I have been engaged in studies of mathematicians' work and, particularly, the study of the relationship between that work and the foundations of mathematics. The central theme of those investigations has been that the problem of the foundations can be formulated as a problem in the production of social order and that so formulated, it can be investigated in and as the work of mathematicians in locally producing, for and among mathematicians, accountably ordinary mathematics. This appendix serves as another introduction to those studies by addressing their possible consequentiality for another problem in the foundations of science: the problem of characterizing the relationship between mathematics and theoretical physics.

The problem of characterizing the relationship between mathematics and physics has its origins in the omnipresence, utility and effectiveness of mathematical formalisms and derivations in the work of theoretical physicists. The question that is usually asked — what is the nature of mathematics and the nature of physics or physical reality that provides for such a mysterious compatibility? — has the effect of turning the real-worldly, material investigation of the mathematical practices of physicists into a problem concerning the naturally and constructively theorized content of the two disciplines.

The aim of this appendix is to indicate that the problem of characterizing the relationship between mathematics and physics, as it is traditionally conceived, is a pseudo-problem; that that problem can be reformulated as a problem in the study of the production of social order:[2] and that that reformulation makes the mathematical methods of physicists capable of being empirically researched as an autonomous praxis.

The idea of my presentation is not to argue each of these claims separately, but to present an argument that provides, at once, for their

APPENDIX

joint intelligibility. Mathematical studies of the so-called foundations of physics have, as a major goal, the development of properly mathematical methods 'substantiating' the mathematical methods actually used by theoretical physicists. I will argue first, that the (theoretically presumed, or programmatically established) existence of such correspondences needs to be separated from the idea that such a correspondence confers a platonic or transcendental character on the mathematical reasoning of physicists, and second, that once this is done, the descriptive adequacy of physicists' mathematical reasoning becomes a curious and interesting phenomenon in its own right.

To this end, I am going to begin by offering a contrast between the mathematical practices of physicists and those of professional mathematicians. A mathematical theorem — the divergence theorem — will be briefly discussed first from the perspective of physicists and then from the perspective of mathematicians. That contrast will set in relief the presupposition of traditional attempts to characterize the relationship between mathematics and physics that the mathematical practices of physicists can be spoken of as, and therein identified with, those of mathematicians. Later in the appendix, I will return to the physicists' proof of the divergence theorem and examine it as a form of instruction in a method of physical reasoning.

A typical statement and proof of the divergence theorem from the perspective of physics instruction is the following: let S be a 'closed surface' and D, the surface S and its interior. Let $\vec{g} = (g^1, g^2, g^3)$ be a vector field defined on D and $\vec{n} = (n^1, n^2, n^3)$, the outward unit normal vector field on S. Then the divergence theorem says

$$\int_S \langle \vec{g}, \vec{n} \rangle dA = \int_D (\nabla \cdot \vec{g}) \, dV.$$

A picture will convey the physical significance of the theorem. In the figure below, let S be the 'closed surface,' let $\vec{g}$ represent the flow of some physical process, like heat, through S, and let $\vec{n}$ be the unit normal vector field on S.

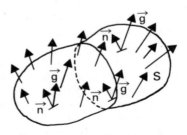

The left-hand side of the divergence theorem is called the flux of the vector field through S and measures, for example, the total heat current directed out from the surface. The divergence theorem says that the

APPENDIX

flux of the vector field through the surface S equals the integral of the divergence of the field taken over the interior of S.

A proof of the theorem can be given in the following fashion.

The first step of the proof is to reduce the problem of proving the theorem to that of proving it for an infinitesimal cube. This is done by showing that the total flux of a vector field through a volume is equal to the sum of the fluxes out of each part of the volume when that volume is dissected into smaller pieces.

The second step of the proof begins by considering an infinitesimal cube.

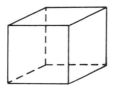

Coordinate axes are arranged so that they line up with the edges of the cube.

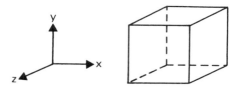

The infinitesimal lengths of the sides of the cube will be denoted dx, dy and dz. In the figure, we also depict the unit normal on one of the faces along with the vector $\vec{g}$ at that point.

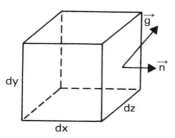

The flux of the vector field $\vec{g}$ through the cube,

$$\int_C \langle \vec{g}, \vec{n} \rangle \, dA =$$

will equal the sum of the fluxes through each face of the cube. Using a

APPENDIX

first-order linear approximation to calculate the flux through two opposing faces, we get

$$\int_C \langle \vec{g}, \vec{n} \rangle \, dA = \frac{\partial g^1}{\partial x} \, dx \, dy \, dz + \ldots$$

and adding similar results obtained from the other sides

$$\int_C \langle \vec{g}, \vec{n} \rangle \, dA = \frac{\partial g^1}{\partial x} \, dx \, dy \, dz + \frac{\partial g^2}{\partial y} \, dx \, dy \, dz + \frac{\partial g^3}{\partial z} \, dx \, dy \, dz.$$

The sum on the right equals the divergence of $\vec{g}$ times the volume of the infinitesimal cube,

$$(\nabla \cdot \vec{g}) \, \Delta V.$$

Using the first step of the proof, we sum these infinitesimal volumes throughout D and arrive at the divergence theorem.

From a standpoint of rigorous mathematics, there are a number of difficulties with this proof, even and especially when that proof is given in greater detail. Although the existence of such difficulties lies at the heart of the study of the mathematical foundations of physics, the particular troubles in the case-at-hand are incidental to the present discussion. What is important is to offer, at least on the basis of immediate visual perception, a direct contrast between the physical version of the theorem and the theorem as seen by professional mathematicians. An actual proof of this theorem would get too involved, but I hope to convey some sense of that proof's technical character.

For mathematicians, the setting of the theorem is immediately generalized to an object of arbitrary finite dimension.[3] Let M be a k-dimensional, compact, oriented C^2 manifold-with-boundary and let ω be a $k-1$ differential form on M. Then the generalized Stokes' theorem says that

$$\int_M d\omega = \int_{\partial M} \omega$$

where ∂m has the induced orientation. This theorem can be proved to the satisfaction of professional mathematicians and, once proved, it can be specialized to the divergence theorem in the case of 3 dimensions. To do this, one defines a 2-dimensional volume element on the tangent space of M

$$dA(v_x, w_x) = \det \begin{pmatrix} v \\ w \\ n(x) \end{pmatrix}.$$

It can be shown that the following relationships hold:

$n^1 \, dA = dy \wedge dz$

$n^2 \, dA = dz \wedge dx$

$n^3 \, dA = dx \wedge dy.$

APPENDIX

Then, by letting

$$\omega = g^1 \, dy \wedge dz + g^2 \, dz \wedge dx + g^3 \, dx \wedge dy,$$

a 2-form on D, and substituting these definitions and relationships into Stokes' theorem, a precise though appearientially similar version of the divergence theorem will result.

This presentation of the divergence theorem illustrates the discrepancies between physicists' mathematical reasoning and that of professional mathematicians. The initial point that I want to make is this: in the reflective discussion of the relationship between mathematics and physics, physicists' mathematical practices are predominantly spoken of as, and therein identified with, those of professional mathematicians. In the case-at-hand, the physicists' proof of the divergence theorem is understood to stand proxy for the more rigorous methods of mathematics. This presumed identifiability of methods underlies the traditionally-conceived problem of characterizing the relationship between the two disciplines.

The obvious descriptive inadequacies of such an identification — illustrated by the example of the divergence theorem and uniformly recognized and acknowledged by mathematicians, physicists, and formal methodologists of science — makes the pervasiveness of its presupposition extremely curious. There are, I believe, two complementary reasons[4] for that pervasiveness, the first of which will be conveyed through another example. At the end of the appendix, I will try to bring these examples together.

In that physicists derive the heat equation from Fourier's law of heat conduction and the law of the conservation of energy (we will write this as follows),

$$\vec{h} = -\kappa \nabla u$$
$$\downarrow \text{ess}$$
$$\Delta u = u_t$$

it is generally supposed that the adequacy of Fourier's law as a description of a physical situation will insure the similar descriptive adequacy of the heat equation.[5] Another way of putting this is that the mathematical derivation (represented by 'the arrow' in the figure above) is presumed to be analytic, or that it is understood to preserve the truth-value of the mathematical descriptions.

In practice, when they are actually engaged in their work, physicists have a much more circumspect regard for their derivations and mathematical descriptions than this account provides. Nevertheless, the account does reflect the way in which physicists will sometimes speak *about* their mathematical practices when, no longer actually at work, they come to speculate reflectively about them. In that their mathematical

APPENDIX

derivations are constructively developed over the course of collegial work sessions and blackboard discussions, those derivations have their retrospectively considered production independence naturally embedded in and as their own communally recognized and accountable efficaciousness.

At this point, I am going to introduce a distinction between the practical analyzability and accountability of physical mathematics as that mathematics is encountered in the work arrangements of theoretical physicists and the problematic character of that mathematical reasoning when it is disengaged from that situated, collegial work. I will refer to the work of that disengagement as the work of an objectifying science. It is from the perspective of an objectifying science that the retrospective efficaciousness of the mathematical reasoning of physicists poses the problem that that efficaciousness could not (at least seemingly) be dependent on the idiosyncracies, circumstantialities, and even dubiously legitimate mathematical methods that pervade the mathematical work of physicists. That such methods actually compose such efficacious derivations is attributed to the existence of rigorous mathematical methods to which physicists' mathematical reasoning alludes and for which it stands as a degraded counterpart.

A second and closely related basis for the thesis of the identifiability of the mathematical methods consists of the unavailability of nonformal methods for technically analyzing the actual use of mathematics by physicists. The known methods by which the reasoning of physicists can be technically and critically analyzed — again from the standpoint of an objectifying science — are those of mathematics proper, and, consequently, to be engaged in the mathematical review of physicists' writings is to already be involved in a reconstructive enterprise that has the identifiability of methods as its programmatic ideal.[6]

In order to summarize all this, let us depict the assumed existence of a platonic mathematical counterpart to physical reasoning with a double arrow so that the initial situation would be mirrored in another situation written as follows:

$$h = -\kappa \nabla u \qquad\qquad h = -\kappa \nabla u$$
$$\downarrow \qquad\qquad\qquad\qquad \Downarrow$$
$$\text{ess} \qquad\qquad\qquad\qquad \text{ess}$$
$$\Delta u = u_t \qquad\qquad\qquad \Delta u = u_t$$

The demonstrated existence of such properly mathematical methods are understood to supply the foundations for the mathematical methods of physicists in that they are seen as providing an account of those methods that is, at least apparently, independent of the actual methods' situated production and use.

Having laid out this material, I am now in a position to come to the main point of the argument. As I indicated earlier and have presented in

APPENDIX

the previous chapters, my investigations of mathematicians' work allow me to evidentially argue that the rigor of mathematicians' reasoning consists of the essentially local work of mathematicians in producing for and among mathematicians the accountably ordinary objects of mathematical discourse. Thus, that the mathematical practices of physicists can be shown, in some sense, to correspond to those of mathematicians (as depicted in the diagram above) can no longer be seen as providing an analytic foundation for physical reasoning. Instead, the situated achievement of some mathematicians in showing that such partial identifications can be made becomes available for what, in fact, it is: the demonstration of the partial translation of one set of practices into another, demonstrated to the satisfaction of one of the groups of practitioners, through which suggestively mathematical methods become rendered as properly mathematical ones.

That the foundations of mathematics, properly conceived, reside in and as the practices that make up mathematicians' work free the question of the relationship between mathematics and physics from the goal of detaching that relationship from its dependence on the situated features of physical theorizing, and return the question to its origins and substantiating conditions in the practice of theoretical physics. Thus, we are taken to the heart of the matter — whether or not it is possible to recover the practical analyzability and accountability of physicists' mathematical methods as, at once, a deeply technical enterprise (as the mathematicians have it) and, also, as an enterprise whose analyzability is essentially available as local, situated accomplishments.

I do not have an answer to that question, but I would like to offer one last example as a way of clarifying it and drawing this material together. Returning to the derivation of the heat equation from Fourier's law, one way of speaking about that derivation (the 'arrow' in the illustration) is to say that it depends centrally on the divergence theorem. A more or less formal rendering of that derivation would make that appear to be the case. Instead, let us look at a more physical derivation.

The flow of heat through a solid is, again, reduced to the problem of the flow of heat through an infinitesimal cube.

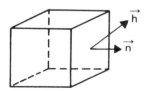

Letting $\vec{g}$ in our previous derivation be replaced with $\vec{h}$, the heat current density, the same reasoning as before gives the total heat current out

APPENDIX

of the cube as the divergence of $\vec{h}$ times the volume of the cube, $(\nabla \cdot \vec{h}) \Delta V$.

As the heat in the object will be proportional to its temperature, the negative of the change of temperature will, by the law of the conservation of energy, equal, up to the constant of proportionality, the loss of heat through the surface of the cube; that is, it will equal the flux. Thus,

$$(\nabla \cdot \vec{h}) \Delta V = \int_C \langle \vec{h}, \vec{n} \rangle \, dA = -\rho \, u_t \, \Delta V,$$

where ρ is the constant of proportionality and the subscript indicates partial differentiation with respect to time. Substituting Fourier's law for the heat current density $\vec{h}$, one obtains

$$-\kappa \nabla \cdot \nabla u = -\rho \, u_t$$

or, essentially,

$$\Delta u \overset{\text{ess}}{=} u_t$$

the heat equation.

What this example seems to do is to make the physicists' proof of the divergence theorem given earlier available as the embodiment of a method of reckoning. The physicist's proof of the divergence theorem serves as instruction in a method of reasoning that, as a method, has the interesting feature that it preserves the physical interpretability of the equations throughout the derivation. The efficacy of the derivation as it concerns physical phenomena does not originate in, nor is it established and maintained by, a correspondence between that derivation and mathematicians' conceptions of proper deductive inference. Instead, it is tied to the constructively and developmentally provided physical interpretability of the equations as that interpretability is constructed and maintained throughout the derivation.

The presence and use of mathematical methods that preserve the interpretability of physical descriptions offers a promising way of initially examining theoretical physicists' actual mathematical reasoning. However, I think it would be wrong to begin looking for instances of structures throughout mathematical physics as straightforward as the example of the divergence theorem, particularly as a means of constructing a documented argument. The relationship between the mathematical formalisms and derivations of physicists and their descriptive content is, inevitably, going to turn out to be extremely subtle in that it must depend on the ways in which a production cohort can make that descriptiveness available through and as the details of their work. What the example does suggest is that the mathematical practices of physicists can be examined independently of mathematicians' practices and that the foundations of mathematical physics might well consist of the concrete, exhibited/exhibitable ways in which that descriptiveness is produced and maintained as a situated accomplishment.

In the philosophy, history and sociology of science, the problem of characterising the relationship between mathematics and physics has a docile presence — that problem is disengaged from considerations of physicists' actual work practices; it is amenable to the endless discourse, historical reviews, and argumentation that make up the practice of those disciplines; it is ignored, commented on, addressed or put aside as a matter of choice. In the work of physicists, where the physical adequacy of mathematical methods is a continual and crucially important concern, that problem takes on a different life entirely. For physicists, the problematic features of the relationship between mathematics and physics are both practically formulated and practically solved as situated, daily features of their work. By examining that relationship as the work of locally producing practically objective mathematical descriptions, the liveliness of that relationship is rediscovered, and the problem of characterizing that relationship is reopened as a problem in the study of the production of social order.

Notes

Introduction

1   The distinction between classical and ethnomethodological studies of work, due to Harold Garfinkel, is a technical one, and I will only elaborate on it here in so far as it applies to the study of mathematicians' work and only in so far as a few brief comments may initially help the reader. The increasingly technical access, as real-worldly researchable matters, that current ethnomethodological studies of mathematicians' work provide to the investigation of the rigor of a mathematical argument as that rigor is essentially tied to the lived-work of that argument's production and exhibition provides as well technical access to the witnessed, yet ignored character of that connection in previous studies of mathematicians' work. In that these previous studies rely on the lived-work of doing rigorous mathematics to find and identify naturally accountable mathematical argumentation — and, thereby, to furnish the adequacy and cogency of their own studies — the essential connection between that lived-work and mathematical rigor is irremediably hidden within those studies in and as the efficaciousness and adequacy of their own practices. Rather than speaking in a non-technical manner of 'traditional' or 'conventional' studies of mathematicians' work, we speak of these studies as 'classical' studies and, thereby, point to and seek to elucidate those studies as themselves practical enterprises tied, for *their* practical accountability, to the various professional disciplines from within which they arise and have their proper and accountable origins. In brief, then, the distinction between classical and ethnomethodological studies of mathematicians' work recommends the existence of classical studies as its own phenomenon in the study of practical action and practical reasoning.

2   As far as possible, given the constraints of the meetings, the intention of my talk was to bring the reader into the presence of

the lived-work of doing mathematics and, by doing so, to offer that work for analysis as practical action. In presenting this material in a lecture, a prover uses his embodied presence to the blackboard and to the audience to achieve the exhibited precision of his work and talk. Thus the {here} noted in the text is replaced or accompanied by a pointing out — by indicating the tracing of — the intercepted arc

where the drawing itself is a temporally developing, temporally organized achievement that can be indicated as follows:

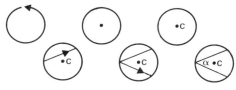

In presenting the talk as a written text, I have retained the use of multiple pictures, arrows and bracketed indexical expressions that initially served the purposes of a prepared lecture to provide instruction for the reader for finding, in the definiteness of the mathematical thing that is then found through their use, the lived-work to which that definiteness is tied. Thus, the reader will not be offered a key to the literary devices being used in the text, but, instead, is advised to seek out in the text's surrounding materials, just what those devices come to, as practical action, in and as the definite thing that they come to exhibit, and to attend to that work as the lived-work of mathematical theorem proving.

3   As an illustration of a 'theory' of such objects, consider, for example, the following definitions of an angle and of its measure (taken from Gustave Choquet, *Geometry in a Modern Setting* (Boston: Houghton Mifflin Company, 1969), p. 79): Let $\mathscr{I}^+$ be the group of even isometries of the Euclidean plane, and $\mathscr{T}$ the normal subgroup of translations. Then an angle can be defined as an element of the factor group

$$\mathscr{A} = \mathscr{I}^+ / \mathscr{T},$$

the angle between a pair of half-lines (A,B) with common vertex as the canonical image in the factor group $\mathscr{I}^+/\mathscr{T}$ of the isometry transforming A into B, and a measure of angles as a continuous

homomorphism φ,

$\varphi : R \to \mathcal{A}$

from the additive group of reals with the natural topology to the angles topologized by first identifying the angles with the multiplicative group of complex numbers of modulus unity. The measure of a given angle α is then defined as the inverse image of α under φ, $m(\alpha) = \varphi^{-1}(\alpha)$. A less exotic example would be the following definition of an inscribed angle: 'Let A and B be distinct points of a circle and let $\widehat{AB}$ be one of the arcs of the chord $\overline{AB}$. We say that an angle is inscribed in this arc if its vertex X is a point of the arc and if its sides are the rays [XA and [XB.' (Howard Levi, *Foundations of Geometry and Trigonometry* (Englewood Cliffs: Prentice-Hall, Inc., 1960), p. 280.)

The reader should note that the notation in the preceding definition speaks on behalf of the work of developing that notation so as to be able to disengage the mathematical objects under consideration from the remarkable relevancies of the work of proving for which that notation was developed as just such a device.

4   The reader may here wish to argue that the need for 'rigorous' definitions of these objects is exhibited from within the local work of the proof. This is certainly the case: as in note 6 below, on entertaining the possibility that

depicts an angle and, thereby, on seeing the possibility of a fourth case involved in proving our theorem, a prover, having in just this way raised the question of what an angle actually is, would find for himself, as the witness to the natural accountability of his own work, the necessity of formulating a definition of an angle which would exhibit that definition's independence from the problematic detail of the proof for which that definition provides a solution. The point to be made is that the need for such a definition and the adequacy of that definition are both, and irremediably, local accomplishments tied to the analyzable detail of an exhibited angle that such a definition is seen to provide and to recover. In this way, then, such definitions not only arise from within, but are answerable to, and must exhibit their necessary properties as part of, the local, lived-work of the proof itself.

5   Let me briefly sketch a different proof of our theorem. (The reader is referred to Levi, *Foundations of Geometry and Trigonometry* for details.) The idea of the proof is to show first that the measure of an inscribed angle depends only on the measure of the intercepted arc. Given this result, the inscribed angles are

NOTES TO PAGES 11-14

partitioned into the class where the intercepted arc is less than 180°, the class where the intercepted arc equals 180°, and the class where the intercepted arc is greater than 180°. For the first case, we need only prove our theorem for the particular case where one of the edges of the angle is a diameter of the circle. The proof would be the same as that of the first case in our original proof. For the second case, we need to prove the theorem only for the situation depicted below:

And in the third case, we can proceed by proving that angles inscribed in opposite arcs of a circle are supplementary

$m(\alpha) + m(\beta) = 180°$

from which case three then follows.

To make this argument work, however, a prover would 'first' ('first' reflecting the locally produced and exhibited orderliness of the work of proving the theorem) have to show that all the angles inscribed in the same arc of a circle are, in fact, congruent to each other, and to do so, a number of other theorems need to be articulated and proved – e.g., one that states that given four distinct points $A_1, \ldots, A_4$ on a circle, exactly two of the chords $\overline{A_1A_2}, \overline{A_1A_3}, \overline{A_1A_4}, \overline{A_2A_3}, \overline{A_2A_4}, \overline{A_3A_4}$, have a point in common and one that states that when the length of two sides of one triangle are proportional to the lengths of two sides of another triangle and, in addition, the corresponding included angles of the triangle are congruent, the triangles are similar.

The existence of this other proof of our proposition allows me to advance another claim: what is involved in proving such a theorem is the articulation of a locally discovered and locally constructed course of proving as a chain of dependent propositions and that, for the mathematician, the heart of his work is not the proving of theorems in themselves, but the mathematical structure building to which that proving provides essential and unique access and, thus, that those structures are themselves local accomplishments or, said differently, that the structures consist of their demonstrable and exhibited properties as built structures and that those structures are 'nothing' other than the organized practices of their construction.

6   As an aside, let me note that a prover, on seeing the possibility of including

193

NOTES TO PAGES 11-28

as an inscribed angle, might not only admit as the representative of an additional equivalence class of inscribed angles, but in proferring the proof of our theorem for this case by defining the measure of so that the theorem holds, a prover might review his proof and come to see the need for explicating the definition of an angle such that would be seen to be the one as well. Moreover, the definition would have to be given so as to preserve the properties of angles and measures of angles exhibited over the course of, and needed for, the proof of our theorem. The import of this aside is not to suggest the transcendental necessity of such a definition; I have already argued that the need for such a definition is itself an achievement of a prover's local work and that the adequacy of such a definition is essentially tied to that work for that adequacy. Instead, the import of this example is that it points to the radically self-organizing (metaphorically, 'hermetic') character of a mathematician's work in producing and exhibiting accountable mathematical structures.

Part I Introduction

1 My use of the word 'rigor' is adopted from conversational usage among mathematicians where it is used to refer to the witnessible (and, thereby, to the apparent, visible, recognizable) adequacy of a proof in demonstrating its own truthfulness, its own objectivity, its own accountability.

Chapter 1 A Review of the Classical Representation of Mathematicians' Work as Formal Logistic Systems

1 This formalization of the axioms of group theory is a slight modification of the one found in Donald W. Barnes and John M. Mack, *An Algebraic Introduction to Mathematical Logic* (New York: Springer-Verlag, Inc., 1975), p. 41.
2 Both the informal axioms for G and the formal ones for G* could be 'weakened,' but such considerations would have greatly complicated our presentation.
3 To the novice, the notion of 'translation' will be seen to obviously trivialize the work involved in being able to 'see' a formal sentence as the translation of a statement of ordinary mathematics.

NOTES TO PAGES 29-42

4 'Elementary' is used here in its usual sense as 'simplest' or 'most basic.'

Chapter 2 An Introduction to Gödel's Incompleteness Theorems

1 'Reasonable' refers to the condition that the set of Gödel numbers for the axioms of the system is a primitive recursive relation.
2 As it is standardly rendered in P.
3 By slightly modifying Gödel's argument, Rosser (1936) was able to show that the first result concerning regular consistency also holds. However, that modification introduces changes in the intuitive interpretation of the undecidable sentence. See F. Feferman, 'Arithmetization of Metamathematics in a General Setting,' *Fundamenta Mathematicae* 49 (1960), pp. 35-92.

Chapter 3 Gödel Numbering and Related Topics

1 Provisional on the complete specification of P.
2 That f is "well-defined" requires proof. I take this occasion to remind the reader explicitly that I do not intend my treatment of Gödel's theorem to itself be 'rigorous,' but to examine, at perspicuous places in a proof of Gödel's theorem, what constitutes the 'rigor' of such a proof. The material reviewed is intended to recall for the reader various features of a proof of Gödel's theorem and to serve as a guide to the work of that proof.
3 E.g., that they are representable in P.
4 E.g., that the representability of the functions $g_1, \ldots, g_n$ and h in the definition of substitution and the representability of g and h in primitive recursion insure the representability of the functions obtained from them by substitution and primitive recursion, respectively.
5 That this result is an immediate consequence of the definitions is exhibited by the following orderly way of proceeding: (i) If $(a_1, \ldots, a_m) \in R$, then $K_R(a_1, \ldots, a_m) = 1$, and from the definition of representability, this implies $\vdash_P K_R(k(a_1), \ldots, k(a_m), S(0))$. (ii) If $(a_1, \ldots, a_m) \notin R$, then $K_P(a_1, \ldots, a_m) = 0$; hence, $\vdash_P K_R(k(a_1), \ldots, k(a_m), 0)$. If $\vdash_P K_R(k(a_1), \ldots, k(a_m), S(0))$, then $\vdash_P 0 = S(0)$ by the uniqueness property (2) of the definition of representability. The axioms of P insure that $\vdash_P 0 \neq S(0)$; hence; by a tautology, $\vdash_P \sim K_R(k(a_1), \ldots, k(a_m), S(0))$. Thus, $R(x_1, \ldots, x_m) = K_R(x_1, \ldots, x_m, S(0))$ numeralwise expresses R in P.
6 The expressions of P consist of all finite, concatenated sequences of primitive symbols of P.
7 The actual practical character of these restrictions will not be available until the material details of the 'structure of proving'

8 Here I adopt the notational convention that $(x, y) \in G$ can be written as $G(x, y)$. The theoretical interchangeability of these notations — available first in the elementary situations where that convention is introduced and, later, through systematic methods of replacing the predicative notation with the set-theoretic one — will be contrasted with the essential character of the collective use of notational conventions for the practical observability of the rigor of the work of proving Gödel's theorem.
9 'Consists' is contained in quotation marks because the identity between the rigor of the proof and these 'demonstrations' is available in and as the 'structure of proving' the theorem. 'Demonstration' is contained in quotation marks because these demonstrations have yet to be explicated as the local work of proving Gödel's theorem.
10 To be discussed later.
11 As opposed to an actual, explicitly displayed formal construction sequence.

Chapter 4 The Double-Diagonalization/'Proof'

1 There is some difference among authors as to what is designated as the double-diagonalization. The sequence of definitions used in constructing the formally undecidable sentence does contain the double-diagonalization of the function $\text{sub}_{g(x_2)}$. The first diagonalization is $\phi(\cdot) = \text{sub}_{g(x_2)}(\cdot, \cdot)$; the 'double-diagonalization' is $\phi(i)$. The apparent fact that $\phi(i) = j$ — apparent in its specific placement within the work for which the sequence of definitions stands as the material display — is sometimes referred to as the 'double-diagonalization trick.' (See below for the definitions of the functions and the numbers i and j that are introduced here.)
2 The stress on this phrase points to the circumstance that the interpretability of these formulas is tied to a way of mathematically proceeding.
3 In writing up the diagonalization, I made use of the treatment of it in Raymond Wilder, *The Foundations of Mathematics*, 2nd edn, (New York: John Wiley and Sons, 1965), pp. 272-3.
4 The selection of x_2 is incidental to the 'proof', but must be consistently maintained throughout the diagonalization and 'proof'. That that selection is incidental is available from reflective consideration of the work of the 'proof'.
5 The practical techniques of working with primitive recursive functions and relations will be addressed in a separate section devoted to them.
6 Historically, knowledge of the sentence *J* as the-sentence-*J*-with-its-undeducibility-properties was instrumental in formulating Gödel's theorem. The 'natural' theorem concerning undeducibility

NOTES TO PAGES 47-53

is 'if P is consistent, then there is a sentence S of P such that neither S nor $\sim S$ are deducible in P.' Gödel's discovery of the sentence J as the sentence with *its* undeducibility properties, and Gödel's ability to formulate that sentence's inadequacy for obtaining the proof of the 'natural' theorem as ω-consistency, allowed Gödel to arrive at the statement of his first incompleteness theorem.

7 As a gloss for the reasoner's work of its facticity.
8 I am borrowing the term 'formating' from Harold Garfinkel. Garfinkel speaks of a formated queue as a queue which exhibits, in and as the positioning of its production cohort, its queue-specific order of service.
9 More generally, to speak of a mathematical discovery is to simultaneously speak of a discovered way of proving.

Chapter 5 A Technical Lemma

1 'Über formal unentscheidbare Sätze der *Principia Mathematica und verwandter Systeme I*,' *Monatschefte für Mathematik und Physik*, 38: 173-98, 1931. An English translation by van Heijenoort is available in Jean van Heijenoort, *From Frege to Gödel: A Source Book in Mathematical Logic 1879-1931* (Cambridge: Harvard University Press, 1967), pp. 596-616.
2 Gödel, in fact, did use the symbol P to designate an (unramified) logistic system similar to that of *Principia Mathematica* together with specific axioms for Peano arithmetic. Gödel's system P corresponds (roughly) to the system PA given below.
3 Alfred North Whitehead and Bertrand Russell, *Principia Mathematica* (Cambridge: Cambridge University Press), Vol. I (1910; second edn, 1925), Vol. II (1912; second edn, 1927), Vol. III (1913; second edn, 1927).
4 Positing the existence of a predicate for every wff.
5 The formulation of the Peano axioms that follows is taken from Chih-Han Sah, *Abstract Algebra*, (New York: Academic Press, 1967), pp. 16-17.
6 I.e., the universal quantifier is relativized by the predicate serving to identify N. The existence of such a predicate N requires the explicit assumption of an axiom of infinity or an equivalent of it.
7 I.e., has the property that if F 'corresponds' in PA to an m-place numercial function f, $\vdash_{PA} \exists! y F(x_1, \ldots, x_m, y)$.
8 Technically, the expression $F(k(a_1), \ldots, k(a_m), k(b))$ would be written in PA as $Z_{a_1}(x_1) \supset \ldots \supset Z_{a_m}(x_m) \supset . Z_b(x_{m+1}) \supset F(x_1, \ldots, x_m, x_{m+1})$, where $Z_n(x)$ is the wff of PA serving to name the number n. The notation used in the text is consistent with Gödel's notation and with the earlier discussion of numeralwise expressibility and representability in P.
9 For the definition of strong representability, the condition of ordinary representability that, for any numbers $a_1, \ldots, a_m$,

197

$\vdash_{PA} \exists!xF(k(a_1), \ldots, k(a_m), x)$ is replaced by the condition $\vdash_{PA} \exists!yF(x_1, \ldots, x_m, y)$. Strong representability implies representability. The idea of strong representability is that the new condition insures that F 'acts' like a function (in PA), and condition (2), the same for representability and strong representability, insures that, when this 'function' is restricted to its appropriate domain, it takes the values that it should.

10 E.g., one whose interpretation in ordinary number theory would be an equation like $5 + 2 = 7$.

11 The notion of the 'length' of a construction of a primitive recursive function will be defined in the next section in terms of a 'formal construction sequence.' Each primitive recursive function can be associated with a 'geneology,' and the induction mentioned above is on the length of that function's 'lineage.'

12 See p. 607 and the accompanying footnotes of the van Heijenoort translation cited above. For the mathematical details of the proof of numeralwise expressibility of primitive recursive relations in PA, the reader is directed to the book by Joel W. Robbin, *Mathematical Logic: A First Course*, (New York: Benjamin/ Cummings Publishing Company, 1969). I take this occasion to express my deep debt to this book and to Herbert Enderton and Louis Narens for helping me to understand the material presented in this section. Responsibility for the text is entirely my own.

13 In principle, the proof of the numeralwise expressibility of primitive recursive relations in PA shows how a wff G numeralwise expressing G can be constructed in PA. The discussion above does not seek to minimize the philosophical import of this fact, but to indicate that the actual construction of G in PA is not essential to the practical objectivity of the proof itself.

14 For the actual construction and proof, see Elliot Mendelson, *Introduction to Mathematical Logic*, (Princeton: D. Van Nostrand, 1964), pp. 131-4.

15 For such a presentation, see Herbert B. Enderton, *A Mathematical Introduction to Logic*, (New York: Academic Press, 1972).

Chapter 6 Primitive Recursive Functions and Relations

1 The case $x < y$ is 'seeably/showably' adequate to the actual condition $x \neq y$. At the end of this chapter, the checks that are actually performed will be elaborated as 'reasoned computations.'

2 As is customary, the designation 'Proof', will be used as the indication that what follows that designation is a conventional, informal proof of the preceding proposition. I.e., 'Proof:' indicates an exhibit of the material detail of a practically accountable proof and not an exegesis or interpretation of the work in which that proof occurs and to which that proof makes reference.

3 The statement of these propositions and their proofs are similar

to those found in Robbin, *Mathematical Logic: A First Course.*
4 I.e., the set of pairs of numbers (x, y) such that x divides y with no remainder.
5 In practice, the primitive recursive equations for a given numerical function or relation are thought of as providing a definition consistent with the function or relation as it is informally and practically known. Thoralf Skolem first indicated how the functions and relations of elementary arithmetic could be defined 'recursively.' ('Begründung der elementaren Arithmatik durch die rekurrierende Denkweise ohne Anwendung scheinbarer Veränderlichen mit unehdlichem Ausdehmungsbereich' ('The Foundations of Elementary Arithmetic Established by means of the Recursive Mode of Thought, without the use of Apparent Variables Ranging over Infinite Domains'), *Videnskapsselskapets skrifter, I. Matematisk-naturvidenskabelig klasse*, no. 6, 1923; translated by Stefan Bauer-Mengelberg and appearing in Van Heijenoort, *From Frege to Gödel*, pp. 302-33). Gödel, in the same paper in which he presented his incompleteness theorem (1931), gave the first precise definition of the notion of (primitive) recursiveness.
6 The set of all primitive symbols, expressions and sequences of expressions of P.
7 The comma usually placed between v_1 and v_2 in $+(v_1 v_2)$ and $\cdot(v_1 v_2)$ is omitted here in that it is used only to increase the legibility of the expression. It is not a primitive symbol of P.
8 I thank Herbert Enderton for helping me work out this formula.

Chapter 7 A Schedule of Proofs

1 On the whole, I have followed Robbin's presentation in *Mathematical Logic: A First Course*, and I have used the notation found there. The reader should note that Robbin proves Gödel's theorem for first-order recursive arithmetic and that changes need to be made to make his proofs adequate for the system P.
2 E.g., if $x = 25 = 5^2 = 2^0 \cdot 3^0 \cdot 5^2 = p_1^0 \cdot p_2^0 \cdot p_3^2$, then $(x)_2 = 0$ and $(x)_3 = 2$.
3 E.g., if $x = 25 = p_1^0 \cdot p_2^0 \cdot p_3^2$, then $L(25) = 3$. L defines the 'length' of a natural number.
4 Roughly, occur(w, x, y, z) indicates that the individual variable with Gödel number x occurs in the wff with Gödel number z.
5 I.e., Sub(x, n, a) equals $g(S_t^x A |)$ or a, where x, t, and a are the Gödel numbers corresponding to x, t, and A, respectively. The reader should note that Sub(x, t, a) does not guarantee that 't is free for x in A.'
6 I.e., sub(x, n, a) = $g(S_{k(n)}^x A |)$ or a, respectively. Since k(n) is a constant, k(n) 'is free for x in A,' and sub(x, n, a) represents a proper substitution without needing the apparatus for specifying

through Gödel numbers the requirement that 't be free for x in A.'

7   This artificiality is heightened by the fact that the numerical relation R need not be the image under a Gödel numbering of any syntactic feature ℜ of P at all. As an illustration of how this may happen, suppose that the Gödel numbers of the primitive symbols of P are all odd numbers greater than 1. Then 4 satisfies the definition of R, but it is not the Gödel number of a symbol, expression, or sequence of expressions.

8   Comparing n with $g(x_n)$, one obtains $0 < 2^3 \cdot 3^9 \cdot 5^5$, $1 < 2^3 \cdot 3^9 \cdot 5^7 \cdot 7^5$, $2 < 2^3 \cdot 3^9 \cdot 5^7 \cdot 7^7 \cdot 11^5$, etc.

9   More elaborately, if $x = g(x_n)$ defines a primitive recursive relation, then $\exists n \leqslant k(x = g(x_n))$ defines a primitive recursive relation of x and k by Proposition 8, and $\exists n \leqslant x \, (x = g(x_n))$ defines a primitive recursive relation of x alone by Proposition 2. However, two points need to be made. First, although this elaboration accounts for the method indicated in the text, the procedure followed in the text is informed by the familiar techniques of working with primitive recursive functions and relations and does not look to such an elaboration for the in-course justification of the construction of $\exists n \leqslant x \, (x = g(x_n))$ as defining a primitive recursive relation. The second point is that speaking of the formula $\exists n \leqslant x \, (x = g(x_n))$ as seeably/showably defining a primitive recursive relation refers to the fact that, given the placement of that formula in the course of work of which the schedule of proofs is its natural account, that formula is itself the materially adequate demonstration that the relation defined by it is primitive recursive and, furthermore, it refers to the fact that the adequacy of that formula is tied to the ability to provide elaborations such as the one just given.

10 That this is so is easy to understand but notationally awkward to prove. If $\alpha_1, \alpha_2, \ldots, \alpha_n$ is an indexed sequence of not necessarily distinct primitive symbols and $a_1, \ldots, a_n$ are the Gödel numbers corresponding to each α_i, the expression $\alpha = \alpha_1 \alpha_2 \ldots \alpha_n$ has Gödel number $2^{a_1} \ldots p_n^{a_n}$. Consider the Gödel number of any proper subset of the α_i, $i = 1, \ldots, n$, concatenated in the order of the increasing index. Each symbol present in α but missing in the new expression represents the deletion of an exponent in the prime factorization of $g(\alpha)$ and the downward shift of the exponents of the larger prime factors. For example, if $\alpha = (S((0)))$, then $g(\alpha) = 2^{\ulcorner(\urcorner} \cdot 3^{\ulcorner S \urcorner} \cdot 5^{\ulcorner(\urcorner} \cdot 7^{\ulcorner(\urcorner} \cdot 11^{\ulcorner 0 \urcorner} \cdot 13^{\ulcorner)\urcorner} \cdot 17^{\ulcorner)\urcorner}$, with 'corners' indicating the Gödel number of the symbol that they enclose. The Gödel number of the expression $\beta = S((0,$ a 'part' of α, is $2^{\ulcorner S \urcorner} \cdot 3^{\ulcorner(\urcorner} \cdot 5^{\ulcorner(\urcorner} \cdot 7^{\ulcorner 0 \urcorner}$. Thus, to obtain the Gödel number of β from the Gödel number of α, the exponent of 2 is deleted in $g(\alpha)$, the other exponents shifted downward, and the two greatest prime factors dropped completely:

$2^{\ulcorner(\urcorner} \cdot 3^{\ulcorner S \urcorner} \cdot 5^{\ulcorner(\urcorner} \cdot 7^{\ulcorner(\urcorner} \cdot 11^{\ulcorner 0 \urcorner} \cdot 13^{\ulcorner)\urcorner} \cdot 17^{\ulcorner)\urcorner} \rightarrow$

NOTES TO PAGES 73-9

$2\ulcorner\lhd\cdot 3\ulcorner S\urcorner\cdot 5\ulcorner(\urcorner\cdot 7\ulcorner(\urcorner\cdot 11\ulcorner 0\urcorner\cdot 13\ulcorner)\urcorner\cdot 17\ulcorner)\urcorner \to$

$2\ulcorner S\urcorner\cdot 3\ulcorner(\urcorner\cdot 5\ulcorner(\urcorner\cdot 7\ulcorner 0\urcorner\cdot 11\ulcorner\rhd\lhd\cdot 13\ulcorner\rhd\lhd \to 2\ulcorner S\urcorner\cdot 3\ulcorner(\urcorner\cdot 5\ulcorner(\urcorner\cdot 7\ulcorner 0\urcorner$.

Each deletion and each shift produces a number smaller than the one before that operation.

The formalization of this reasoning is uninstructive and is omitted here. In fact, even the elaboration that has been given is never presented in a proof of Gödel's theorem: the assertion that x is a bound on v is understood as a sufficient basis for the reconstruction of this reasoning and, therein, the uncovering of one more aspect of the structure of the Gödel numbering and its compatibility with the techniques of working with primitive recursive functions and relations.

11 The choice of t is not completely arbitrary in that it has been constructed so as to include all the term building operations of P. In that this construction is intentional, results concerning t particularly are seen to be capable of generalization to the entire class of terms of P.

12 In that $x = g(0)$ and var(x) have been shown to define primitive recursive relations independently of the construction of the formula for term(x), the presence of them in the above formula is unproblematic.

13 I.e., as part of the arranging of practices that makes up the search for a bound on y.

14 Cf. note 10.

15 The circumstance that $L(x) = 0$ for $x = 0$ and $x = 1$ suggests the need to test the formula for these cases.

16 By definition, G(x, u) is equivalent to ded(x, sub(g(x_2), u, u)). If sub(x, n, a) defines a primitive recursive function, then sub (g(x_2), n, a) defines a primitive recursive function by Proposition 2 (the substitution of a constant for a variable in a primitive recursive function), and by Proposition 2 again (identification of variables), sub(g(x_1), u, u) defines one, also. Thus, if ded(x, y) defines a primitive recursive relation, G is primitive recursive by Proposition 7 (the substitution of primitive recursive function for a variable in a primitive recursive relation).

17 This proof and several of those that follow have been taken from Joel W. Robbin, *Mathematical Logic: A First Course* (New York: Benjamin/Cummings Publishing Company, 1969).

18 Or *transformation rules*.

19 The placement of the definitions concerning the operation of substitution — that is, whether they are part of the formation rules as 'grammatical' entities or are part of the axiomatics — is somewhat problematic for the abstract definition of a formal system. The important point here is that in practice, for a particular formal system, the terms 'formation rules' and 'axiomatics' are understood to refer to a definite and semi-ordered collection of syntactic definitions. Definiteness of reference has

been supplied above by fiat in order to avoid a potentially interesting but, also, a potentially distracting discussion of the vernacular character of the recognizable definiteness of such definitions (including that of the syntax of a formal system). I thank Harold Garfinkel for bringing this matter to my attention and for offering his reflections on it.

20 Or 'logical syntax'.

21 The first person singular pronoun is assuredly not being used to indicate originality of conception, but only to emphasize (1) the attribution of a referential definiteness to the notion of the syntax of a formal system and (2) the fact that the definition of syntax that is given refers to syntactic objects and not to their definitions or, alternatively, that this definition emphasizes the strictly extensional character of the syntactic definitions that are involved.

22 That is, the primitive symbols, the expressions, and the sequence of expressions.

23 Theorem(x) is defined as holding if and only if x is the Gödel number of a *theorem* of P.

24 The faulted character of this definition will be discussed later in the text, and an adequate definition will be given.

25 'Associativity' refers to the property that, for all x, y, and z, $(x * y) * z = x * (y * z)$. Because of this property, the 'product' $x * y * z$ is unambiguous. The importance of the associativity of * will be indicated later in the text.

26 Empty parentheses are used in the propositions shown later in the text to indicate that although those propositions are serially ordered on the page of working notes or are indexed on that page as an ordered sequence, they are not yet specifically placed in the increasingly articulated, finished schedule.

27 Read: 'x divides y if and only if there exists a natural number n less than y such that $y = n \cdot x$.'

28 Read: 'x is a prime number if and only if x is greater than 1 (i.e., $x \neq 0$ and $x \neq 1$) and for all numbers y less than x, if y divides x then y is either 1 or x.'

29 Read: 'The (n + 1)-th prime is the least number x (less than or equal to $\square$) such that x is prime and x is greater than the n-th prime.' In order to use the μ- or least number operator, an upper bound on x must be found. The 'empty' square is a circumstantially-occasioned, situationally-transparent device that a prover could use to signal the need for such a bound without being distracted from the work of finding the main body of the formula.

30 For instance, p_0 could have been defined earlier as 2 as is done in Stephan Cole Kleene, *Introduction to Metamathematics* (New York: D. Van Nostrand Inc., 1952).

31 The arbitrariness of s comes from the fact that $p_0 = 1^s = 1$ for all s.

32 The arbitrariness of r comes from the fact that $p_r^0 = 1$ for all r.

33 The notion that this solution appears as a 'natural' one can be elaborated by pointing to the way in which that solution is embedded in the surrounding work of developing that formula; the 'naturalness' of the solution is tied to the fact that x no longer needs to be decomposed into its prime factorization, that the formula applies when x is either 0 or 1, that x * y will be primitive recursive if the remainder of the formula can be constructed so as to exhibit its primitive recursiveness, that the formula portends the use of the finite product construction
$$x \cdot p_0^0 \cdot \prod_{i=0}^{n} p_{m+i}^{b_i}$$ as indicated immediately below in the text, and, finally, in that all this is so, that this solution, when it is found, already fits within and further develops the immediate course of work that seeks an adequate definition of * and that provides the background for what an adequate definition of it could be.

34 Lg(P) was defined previously as the 'language' of P — i.e., as the set of all primitive symbols, expressions, and sequences of expressions of P.

35 I.e., by using the technique of constructing primitive recursive functions that appears as Proposition 9 in the schedule of proofs outlined earlier.

36 Or knows as a possibility.

37 Given an initial way of working similar to the one depicted above.

38 In the sense in which such a 'determination' is given later in the text.

39 I.e., the inverse image under the mapping $n \mapsto p_n$.

40 For that matter, the prover may examine a familiar text or texts to find, through that consultation, either initially or exactly how to proceed. An elaboration of this issue will be given in points (iv) and (v) of the summary of the topic (a).

41 The idea of speaking of the 'co-temporaneous' use of both L and ( ). in writing the formula for x * y is that the use of both of these functions may arise while working on writing the 'final' equation (i.e., the equation that is found to need no further modification throughout the rest of the work of developing the schedule), but that L and ( ). are neither simultaneously invoked nor 'remembered' in a necessarily temporally-ordered fashion. 'Co-temporaneous' refers to the fact that the need and use of these functions may arise in the presence of one another and as the 'working-at' and 'working out' of the same immediate problem.

42 Read: 'The "length" of x, L(x), is the number m of the least prime p_m such that p_m divides x but p_k does not divide x for all $k > m$.' The upper bounds on m and k need to be large enough to insure the adequacy of the formula, but since $x < p_x$ for all x, x itself will serve as these bounds.

43 That L(0) and L(1) 'should' have these values is a reference not

only to the fact that 0 and 1 have no conventionally understood prime factorizations, but also to the way in which the values of L at 0 and 1 'fit' into the schedule of proofs as a developing system of local relevancies. In the case of L, this 'fit' is illustrated by the compatibility of $L(1) = 0$ with writing 1 as p_0^0, and it will be illustrated further later in the text; the more general point will be elaborated when I come to summarize some of the material in topic (a) by speaking of the 'intrinsic' orderliness of the schedule of proofs (point (vi)).

44 Read: 'The exponent of the i-th prime in the prime factorization of x is the least number k (less than x) such that p_i^k divides x but p_i^{k+1} does not.'

45 This is technically incorrect. In the cases mentioned, the μ-operator is defined (by the convention indicated in the schedule of proofs outlined on pages 65–9) to equal 0, and this is exactly the value that will be explicitly stipulated for ( )$_.$ in the text below. In the list that follows, I give possible 'reasons' for the prover, as he is developing the schedule, to find the 'inadequacy' of the definition of ( )$_.$ as that 'inadequacy' is noted above. In speaking of them as possible 'reasons,' I do not intend them as an exhaustive list of distinct cases, but as an attempt to provide for the indefinite number of ways a prover may come to develop an always this-particular schedule of proofs. Of course, it remains that a particular prover may not find this specific matter problematic, and this, too, as being tied to *his* way of developing, in its material specificity, a schedule of proofs. 'His way,' however, would be capable of similar elaborations. The list is as follows: (1) Although the prover is using the μ-operator when he is writing the formula for ( )$_.$, he may not have as-of-yet explicitly worked out its definition. (2) In working with a half 'verbal' translation of the μ-operator, the prover may find through that translation — in that that 'translation' does not make available the precise definition of the μ-operator — the inadequacy of the definition of ( )$_.$. (3) The fact that ( )$_.$ is well-defined by the equation given in the text does not mean that the definition given by the μ-operator is appropriate to the case at hand, and the prover may find the need to explicitly work this out. (4) The prover may explicitly define ( )$_.$ when $x = 0$ and $i = 0$ as a means of making explicit just what is needed for the local contingencies of the work of producing the schedule of proofs. (5) In developing his schedule, the prover may be consulting other texts giving such schedules of proofs, and he may see in those texts (a) the lack of such a specification and, although by working back through the definitions, he may find ( )$_.$ to be adequately defined, he may nevertheless feel that the accountable work of the schedule's production has not been made sufficiently available; (b) the lack of such a specification and, by incorrectly finding ( )$_.$ as not being well-defined, thereby finding as well his own contribution to the previous provers'

work by correcting that definition; or (c) in that the writers of those texts may have used other conventions in developing their schedules of proofs, the prover may find himself with the continual task of appropriately translating the other schedules into his own current way of working, and the apparent need to define ( )$_\bullet$ for $x = 0$ and $i = 0$ may arise in and as the circumstances of such translational work.

46 In his seminars, in speaking about the work of a scientific discovery, Harold Garfinkel suggested and elaborated on the reinterpretation of varous themes from Gestalt psychology — like that of the figure-ground relation and that of the gestalt-switch — in terms of the situated and endogenously developing practices of finding and exhibiting a scientific discovery. Rather than understanding, for example, the 'ground' as the current state of (physical, astronomical, chemical, etc.) knowledge and experimentation concerning the domain of phenomena against which the discovery is then referred, and the 'figure' as (what amounts to a statement asserting the existence of) the discovered object itself, Garfinkel suggested that the 'ground' be interpreted as the situated, known, and recognized practices making up adequate and efficacious laboratory procedure and that the 'figure' be interpreted as the cultivated and discovered organization of those practices to find again the thereby naturally accountable and propertied object. By this note I do not mean to attempt to summarize Garfinkel's work on this matter, but to point out that my reference to the 'discovered gestalt' of the material indicated in the text is a reference to the discovered coherence of the definitions that are given against the background of the surrounding practices of finding those definitions, and that that coherence consists of the ways in which, when once found, the propositions, their proofs, and the associated definitions seeably/showably elaborate one another as making up an accountable course of proving. The reader is referred to Harold Garfinkel, Michael Lynch, and Eric Livingston, 'The Work of the Discovering Sciences Construed with Materials from the Optically Discovered Pulsar,' *Philosophy of the Social Sciences* 11 (1981), pp. 131-58.

47 This qualification is being used to signal (1) that this formulation of the prover's use of the *-operator stands *in relation to* the actual work practices that make up that use and, therefore, as such a formulation, can be examined for its faulted character, and (2) that this is so is not crucial for the discussion that follows.

48 In distinction from N, $g(Lg(P))$ being a proper subset of N.

49 Where 'examine' is being used as a gloss for the work of using and interrogating the formula with the intention of determining whether or not, or the conditions under which, * is associative.

50 Two alternatives to the course of reasoning that I have sketched should be mentioned explicitly. In that the prover may only come to see the need to assert the associative property of * when he finds that he needs that property for proving later propositions

of the schedule, that assertion and an appropriate qualification on the domain of its validity may occur as retrospective modifications on the previously established formula $x * y = x \cdot L(y) \cdot \prod_{i=0}^{L(y)} p_{L(x)+i}^{(y)_i}$. On the other hand, the prover may be constructing his schedule of proofs in close consultation with texts providing such schedules, and the prover may adopt the device used above from those texts in that he comes to see, through that consultation, the utility of it.

51 Even though only the case $y = 0$ creates problems for the associativity of *, the stated restriction on that associativity is that x, y and z are all non-zero. That a prover will write the later statement as the restriction and not the more specific one of $y \neq 0$ (or the even more precise condition that $y \neq 0$ or that $y = 0$ and either $x = 0$ or $z = 0$, also) points to the following phenomenon: a prover will know, as part of the background of work against which he is writing and organizing the accountable schedule of proofs, the details of the schedule's reasoning that he is not making explicit; he will recognize that certain statements can be made more precise and that certain arguments can be elaborated, but he will not attend to such precision or provide such elaborations. It is not, however that the prover is leaving out *some* of the details of the proof, for the associated quantifier — *all* of the details of the proof — does not have a clear referent. Instead, a mathematician reviewing a proof will point to the omission of details that he considers as *necessary* in providing for the accountable work of which the written proof is a residue and for which the written proof serves as a guide. The prover, in turn, is responsive to the problem of setting out those details of a proof that are identifying of that proof as a naturally accountable proof of the theorem in question. In the case at hand, in writing the restriction on x, y and z as the restriction that they all be greater than 0, the prover is speaking faithfully of the witnessed details of the work of proving Gödel's theorem while, at the same time, not holding himself to the endless elaboration of them. In that a proof is written and read as a pedagogical object — in that it is intended as and is held accountable for making available or 'teaching' a way of proving — a proof invites the reader into reworking the proof to find how the written proof does, in fact, stand as the accountable work of proving the theorem in question. Once again, in the case at hand, the prover leaves it to the reader to see that the condition on the associativity of * is both adequate and too restrictive, and, in seeing that it is too restrictive, that a concern over this precision is of no consequence to the accountable production of the schedule of proofs.

52 Parenthetically, in developing the presentation given here, I initially wrote the equation for $\langle a_1, \ldots, a_m \rangle$ without exhibiting

NOTES TO PAGE 92

the i-th term $p_{Z_i^m}^{I_i^m}$. Having written that equation and having then come to see what could be done — as a prospective course of action within the course of the ongoing analysis — to provide further a structure of construction for $\langle a_1, \ldots, a_m \rangle$, I then introduced the term $p_{Z_i^m}^{I_i^m}$ which took on its relevance in developing the analysis of the next note but one by making notationally available what the i-th term of the product would be. The point that I wish to make is that this is just one of an endless number of examples of the way in which the specific, material detail of a written course of accountable mathematical proving is cultivated over the course of that proof's production. (I am indebted to Herbert Enderton for pointing out that, in that the functions $\langle a_1, \ldots, a_m \rangle$ are defined for each m separately, there is no need to indicate a dependence of any one of those functions on the function $n \mapsto p_n$ enumerating the prime numbers. Thus, for example $\langle a_1, a_2 \rangle = 2^{a_1} \cdot 3^{a_2} = (Z_2^2(a_1, a_2))^{a_1} \cdot (Z_3^2(a_1, a_2))^{a_2}$. Moreover, given this analysis, the arrows from p_n to $\langle x_1, \ldots, x_n \rangle$ on page 96 and from 13 to $\langle x_1, \ldots, x_n \rangle$ on page 97 can be eliminated. The analysis that is given in the text was based on the fact that to write a general formula for $\langle a_1, \ldots, a_m \rangle$, a prover 'will' write $\langle a_1, \ldots, a_m \rangle = p_1^{a_1} \cdot \ldots \cdot p_m^{a_m}$, and given this circumstance, a prover 'might' — as I did — proceed to obtain the constants p_i by composing the function p_n with that of Z_i^m, thereby arriving at (what is now available as) the somewhat 'erroneous' construction of $\langle a_1, \ldots, a_m \rangle$ that appears in the text and these accompanying notes. The major point of the text and notes should, however, be remembered: for a prover engaged in producing or reviewing a schedule of proofs, the primitive recursiveness of $\langle a_1, \ldots, a_m \rangle$ does not depend on that prover actually working out an explicit construction of $\langle a_1, \ldots, a_m \rangle$ from the definition of a primitive recursive function, but on his seeing a way of showing that it can be so constructed. Later, in the chapter 'A Structure of Proving,' I point out that the relevant detail of a proof is tied to the structure of that proof's local work and that, in this way, what is available at the mathematical work-site as the relevant detail of a proof can be different from within different work circumstances. This does not mean that 'anything goes,' but that 'what goes' and 'what is correct' are themselves, in each particular case, tied to, and discovered as, local work practices. This technical aside does not affect the argument in the text. I am grateful to Herbert Enderton for his assistance; responsibility for the text, however, is mine alone.)

53 Whose unremarkable character (and in consequence, the somewhat exaggerated use of the word 'innovation') is connected not only to that notation's intended/discoverable meaning, but, as

207

NOTES TO PAGES 92-3

well, to the presence of the similar use of the superscript in the notation for the projection functions I_i^m.

54   The finite product $(a_1, \ldots, a_m) \mapsto a_1 \cdot \ldots \cdot a_m$ is primitive recursive in that it itself is seeably/showably constructed by the substitution of a previously constructed primitive recursive function $(a_1, \ldots, a_{m-1}) \mapsto a_1 \cdot \ldots \cdot a_{m-1}$ for a variable in the 2-place primitive recursive function of multiplication and such a construction requires only a finite number (in this case, $m-1$) of steps. Exhibiting that substitution through the formula

$$(a_1, \ldots, a_m) \mapsto \left(\prod_{i=0}^{m-1} a_i \right) \cdot a_m$$

simultaneously indicates how an inductive proof could be given.

55   In even greater detail, by using the notation $ex(a, b)$ for a^b, $\langle a_1, \ldots, a_m \rangle$ can be analyzed as being constructed as

$$(a_1, \ldots, a_m) \mapsto (ex(pz_1^m{}_{(a_1,\ldots,a_m)}, I_1^m(a_1, \ldots, a_m)), \ldots,$$
$$ex(pz_m^m{}_{(a_1,\ldots,a_m)}, I_m^m(a_1, \ldots, a_m)))$$
$$\mapsto \left\{ \prod_{i=0}^{m-1} I_i^m(ex(p_1, I_1^m(a_1, \ldots, a_m)), \ldots, ex(p_m, I_m^m(a_1, \ldots, a_m))) \right\} \cdot I_m^m(ex(p_1, I_1^m(a_1, \ldots, a_m)), \ldots, ex(p_m, I_m^m(a_1, \ldots, a_m))) = p_1^{a_1} \cdot \ldots \cdot p_m^{a_m}.$$

56   Given the way of working from within which those propositions and their proofs have been constructed. For further amplification of this point, see the summarizing material (i)–(vi) on pages 94–100.

57   The propositions and proofs that follow in the text are presented as they might appear in a finished schedule of proofs. In constrast, the prover who has developed the schedule to this point will (probably) only have those propositions and proofs enunciated and arranged as, for example:

— $x|y$ is p.r.

   Pf: $x|y \Leftrightarrow x \neq 0 \land \exists n \leq x \, (y = n \cdot x)$

— $prime(x)$ is p.r.

   Pf: $prime(x) \Leftrightarrow (x > 1) \land \forall y \leq x \, \{ y|x \supset (y = 1 \lor y = x) \}$

— $n \mapsto p_n$ is p.r.

   .
   .
   .

— $\langle a_1^{x_1}, \ldots, a_m^{x_m} \rangle = p_1^{x_1} \cdot \ldots \cdot p_m^{x_m}$

   Pf: by ~~definition~~ construction

In fact, the physical ordering of the propositions on the page might be different; (possibly) arrows or sequential numberings superimposed on one another (as provers' situationally-occasioned devices) might be used, or propositions and proofs might be sandwiched between other propositions and proofs on the working page, so as to mark the accountable reorganization of those propositions and proofs and, therein, the accountable orderliness of the prover's work as well. The surrounding way of speaking about those propositions and proofs — that is, whether or not, or to what degree, the distinction between the syntax language, the meta-language, and the language of ordinary number theory should be applied (e.g., whether to use $\supset$ or 'implies' in the proofs of propositions given in ordinary number theory), whether or not such distinctions should be specifically mentioned and used in the text, whether the 'proofs' of the propositions should be separated from the propositions and annotated as 'proofs' or, instead, be given as definitions with accompanying 'verbal' interpretations (as Gödel did in his original paper), whether functions such as $n \mapsto p_n$ should be written in this fashion or abbreviated as, in this case, p_n, whether motivational remarks and/or suggestive names should accompany the introduction of the 'apparatus' ( )., L, *, and $\langle \rangle$, or, instead, the reader should be left to come into that 'motivation' in and as the further work of proving the propositions of the schedule, etc., etc. — may not have been 'decided' or 'finalized' as of yet, and the prover might postpone, or may not have yet come into the relevance of, working out such details for the 'final' writing up of the proof. ('Decided' is placed within quotation marks in that, once again, the term is being used to refer to the practices that make up the writing of the schedule as those practices are tied to the material detail of the schedule itself, and not, as might be imagined, to refer to a psychological process that can be disengaged from those practices and that detail.)

58 Although the initial presentation of the 'encoding' functions $\langle \rangle$ may have suggested the centrality of those functions for the construction of the schedule of proofs, their utility is hindered by the fact that a separate function is needed for each number of arguments. In that the Gödel numbering already 'encodes' sequences of numbers in single numbers, much of the work of producing the schedule can be formulated as the work of 'decoding' a given number so as to exhibit the 'structure' of its construction. For the formal theory P, the prover does not need to use any of the functions $\langle \rangle$ as a prover would discover over the course of his working through the entire schedule. In the 'final' version of the schedule, the proposition concerning $\langle \rangle$ would be omitted. The reader will find an example of the use of $\langle \rangle$ in a schedule of proofs in Joel W. Robbin, *Mathematical Logic: A First Course* (New York: Benjamin/Cummings Publishing Company, 1969), pp. 103-4.

59 Or to have retrospectively introduced on the occasion of their apparent need, or to have begun the construction of the schedule by first laying out the following functions as being prospectively

60. Respectively, the characteristic function of the set $\{0\}$, the characteristic function of the set $N - \{0\}$, and the absolute value of the difference of two numbers. $\overline{sg}(x)$ is defined by primitive recursion $-\overline{sg}(0) = 0$, $sg(S(x)) = 1$ — and $sg(x)$ is then defined as $1 \mathbin{\dot{-}} \overline{sg}(x)$. The reader will recall that $x \mathbin{\dot{-}} y$, equaling $x - y$ if $x \geqslant y$ and 0 otherwise, was introduced earlier as a primitive recursive function.

61. Where 'similarly' refers to the ways in which the introduction of that function is already embedded in the surrounding work of proving that provides for that function's (prospectively) efficacious use.

62. Where the appropriateness of using 'then' is now to be understood as having its origins within the achieved availability of the accountable work of the schedule's production.

63. A sequence of proofs similar to the one that follows can be found in Mendelson, *Introduction to Mathematical Logic*. The reader should note that Mendelson defines the characteristic function on a subset A of N^m as being 0 on A and 1 on $N^m - A$, and that I have followed the more standard mathematical practice (if not the standard practice of mathematical logicians) in defining the characteristic function as 1 on A and as 0 on $N^m - A$.

64. To speak of the produced schedule as a mathematical discovery may seem to the reader a usage more properly reserved for more significant 'results' as, for example, the proof of Gödel's theorem as a whole. The point, however, is that this use of 'mathematical discovery' does not concern the importance of a result, but speaks instead of the lived-work of proving that inhabits and is identifying of mathematical practice.

65. If, however, the graph included the first ten propositions of the schedule, the number of lines would obscure the immediate availability of the thing that such a representation proposes. Similarly, for the prover constructing such a graph, a central theme and a recurring problem of that construction is the arrangement of the graph so as to exhibit the organizational features of the schedule that that graph is intended to represent.

66. I am indebted to Harold Garfinkel for the notion that such a graph 'renders' (the schedule of proofs/the lived-work of the schedule's production that is identifying of the schedule as such) as a 'structure of logical dependencies.'

67. Again, I am indebted to Harold Garfinkel for making the notions of a 'received view,' 'received history,' and 'received topic' available as a matter of ethnomethodological interest.

68. The word 'might' is used here to indicate that the inquiry that is embodied in the work of composing such a derivation is itself part of, and arises from within, a developing, material-specific way of working through a schedule of proofs; the usage is not

intended to imply that a prover 'chooses' between 'subjectively' available, alternative courses of reasoning and action.

69 Although the use of the word 'substitution' can be justified in terms of the relation defined by the composition of the characteristic function of the equality relation, $K_=(I_1^2, I_2^2)$ with (I_1^3, I_2^3, I_3^3), the definiteness of the reference to "the relation that results from "substitution"' (as with the definiteness of reference provided by the use of 'identification' later in the text) does not reside in its exegesis as an accountable description, but is first provided for as a definiteness of action in writing the derivation itself — specifically, the equation $y = n \cdot x$ is written below (and 'thereby' follows from) the equation $y = z$ with $n \cdot x$ (as a purely symbolic or 'formal' entity) being seen as replacing, or as being substituted for, z. That it is through such definiteness of action that the statements of the initial propositions (in this case, Proposition 7) take on their practical utility serves as a reminder that the reasoning given in this paragraph stands only *in relation to* and *renders* the combined course of writing and reasoning that *makes up* and *is* the derivation itself.

70 In greater detail, Proposition 2 permits the identification of variables y and w in the characteristic function of the relation defined by $\exists n \leq w \, (y = n \cdot x)$. Alternatively, in that y (viewed as a projection function) is itself primitive recursive, the primitive recursiveness of $\exists n \leq y \, (y = n \cdot x)$ follows from that of $\exists n \leq w \, (y = n \cdot x)$ by Proposition 7, the substitution of a primitive recursive function for a variable in a primitive recursive relation. The relationship between Proposition 2 and Proposition 7 will be developed later in the text.

71 It might be argued that a prover is aware, over the course of writing/verbalizing the displayed formula, that that formula and verbalization are practical expedients whose approximate character can be made available, for example, by (interpretatively) amplifying the given definition so as to read: *if x and y are natural numbers* (repaired in writing/verbalizing-to-oneself to read 'for all x and y natural numbers'), *then* x *is said to* divide y if and only if $x \neq 0$ and there exists a *natural* number n such that y equals n time x:

$\forall x, y \in N, x|y \Leftrightarrow x \neq 0$ and $\exists n \, (y = n \cdot x)$.

In this way, the earlier definition comes to be seen as modeling, in known and analyzable ways, the 'real' properties and objects that it represents and to which it refers. A first point to be made about such an argument is that a prover neither uses nor relies upon such exactitude of expression in order to write the formula

$x|y \Leftrightarrow x \neq 0$ and $\exists n \, (y = n \cdot x)$.

as, at that time, the practically adequate definition of the divisibility relation. Secondly, if incorporated into a schedule of proofs, such exactitude would only obscure the developing orderliness of a prover's work. But the larger point is this: whether or not

$$x|y \Leftrightarrow x \neq 0 \text{ and } \exists n \, (y = n \cdot x)$$

is recognized by the prover, as he is developing the schedule, as being potentially remediably-vague, the precision that may come to be required of the written formula, as such a requirement may become motivated by the detailed circumstances of the prover's immediate work, is itself required to be available in and as the accountably-historicized, exhibitably-relevant material detail of that work itself. In this way, the accountable corrigibility or incorrigibility of the definition is essentially tied to that definition's placement within an accountable course of proving's work. A prover, on later coming to see the need to assert that the definition refers to *natural numbers* x and y, will extract from his course of work — as an historicized, material-specific object — the reasoning and detail needed to make that fact apparent (and, in this particular case, utterly 'trivial').

72 The use of 'provisionally' needs some amplification. First, the term is somewhat misleading: although it appears to be descriptive, it glosses the ways in which such a formula is, for a prover, part of, and remediably available as part of, an ongoing course of work. On the other hand, in that this term signals the utterly practical character of a prover's work in constructing a schedule of proofs, it reminds us, once again, that the accountability and analyzability of the schedule is a practical achievement and, thus, recommends, once again, the work of that achievement as the phenomenon of interest.

73 The quotation marks around 'defined' are used here to indicate that at this point in developing a schedule of proofs, a prover 'might' not (in the sense of note 68) thereupon explicitly formulate a definition of the relation in question.

74 Both in terms of the particular relations involved and in terms of the notational specifics in which those formulas are written.

75 Specifically, a prover 'may' distinguish the cases when $y = 0$ and when $y > 0$. If $y = 0$, then $y = 0 \cdot x$ for all x and $x|y$ for all x with $n = 0$. Thus, if $y = 0$, $x|y \Leftrightarrow \exists n \leq y \, (y = n \cdot x)$. On the other hand, if $y > 0$ and $y = n \cdot x$ (i.e., $x|y$), then neither n nor x can be equal to 0. An inductive proof shows that for all $x, y > 0$, $y \leq y \cdot x$. Hence, y is again an upper bound on n.

The reader should note that it is unlikely that an actual prover would articulate his reasoning to this extent. Instead, a prover would probably see — perhaps through a review of a few particular cases examined as exemplars — that the distinction between the cases when $y = 0$ and when $y > 0$ seeably/showably provides for a line of argumentation similar to the one just given, and he would anticipate both the theoretical simplicity and picayune detail of materially elaborating that argument and, thereby, of further articulating and exhibiting its reasoning.

NOTES TO PAGES 104-6

76 To speak of the following way of working as a 'method' is not to thereby claim that it consists of a definite set of procedures and abbreviatory practices. Instead, the methodic character of this way of working is itself increasingly developed and articulated over the course of the prover's work, thereby taking on expressive capabilities and analytic properties not provided by its summarizing description. In fact, the adequacy of the description of that method is available to a prover only after having come to see how its methodic character is tied to and developed as both a way of working and, specifically and crucially, as a way of working that has, as its accomplishment, the adequate demonstration of the adequacy of the proofs in question.

77 The reader should note that although the availability of the technique of bounded quantification was used in writing the displayed equations and, for the prover, is part of that display as a derivation, a prover may not have seen as of yet the pointedness of incorporating that use in the material display itself.

78 One way of determining whether or not Proposition 8 could be used directly — and, thus, an alternative procedure to that developed later in the text — would be to review the various constructions that are used in proving Proposition 8 to see if the particular aspect of its notation that is in question reflects a substantive matter or is, instead, merely an inadvertent result of the particular formalism in which that proposition was written. By working backward through a proof of Proposition 8, a prover might find (as his practical achievement) that, at some stage in the construction of that proof, a technique like the one formulated in Proposition 2 was 'already' used; therein, as a seeable/showable consequence of that finding, the prover will have found the accountable irrelevance of a new variable for the upper bound on n.

79 At this point in constructing the derivation, a prover might realize only that what is involved is an 'identification' of variables and that Proposition 2 permits such identifications. If this were so, it would remain for the prover's retrospective review of his own work to clarify the fact that the substitution in question is actually being made in $K_R(w, y, x)$, the characteristic function of the relation $R(w, y, x)$, and not in $R(w, y, x)$ itself, thereby formulating the immediate problem as a proper instance of Proposition 2.

80 If a prover were to have adopted a different method of working out such a derivation — for example, by associating a relation symbol with the relation that an equation defines:

$T(y, n, x) \Leftrightarrow y = n \cdot x$ ( )

$S(y, w, x) \Leftrightarrow \exists n \leqslant w \, (T(y, n, x)) \Leftrightarrow \exists n \leqslant w \, (y = n \cdot x)$ (8)

$R(y, x) \Leftrightarrow S(y, y, x) \Leftrightarrow \exists n \leqslant y \, (y = n \cdot x)$ (2)

— then this new method would have raised problems for the

derivation that were not exhibited by the method described in the text. An example of such a problem is the fact that the relation x|y 'transposes' the variables in R(y, x), a course of writing again provided for by Proposition 2 of the schedule outlined earlier. In practice, of course, a prover will not hold himself to any one 'method,' but will use those devices that become relevant to the developing material detail of the particular problem-at-hand as, in fact, was done in the text above. Over the course of constructing such a derivation, a prover will, as part of that work itself, do that work in such a way that it could be articulated as an accountable definite procedure; further, he will look to that work's prospective achievement in finding and exhibiting such a derivation in order to retrospectively insure — as a rehearsed course of writing and reasoning — the accountability of that work itself. At the same time it should be noted that a prover will find no immediate, material motivation for his actually articulating as a definite procedure the situated way of working that he is both using and developing.

81 'May not give' is used here in preference to 'may fail to give,' the point being that the latter expression preserves the transcendental necessity of such a review whereas the former is consonant with our examination of the relevant and problematic details of such formulas as those details are tied to and embedded in the course of writing/reasoning of which they themselves are a part.

82 The condition '$x \neq 0$', not considered in the discussion immediately following in the text, will be briefly examined at the end of topic (b). However, for completeness, let us note the following: first, if both '$x \neq 0$' and '$\exists n \leq y \, (y = n \cdot x)$' define primitive recursive relations, then so does '$x \neq 0$ and $\exists n \leq y \, (y = n \cdot x)$' in that definitions formed through the use of sentential connectives (i.e., 'not,' 'and,' 'or,' 'implies,' and 'if and only if') preserve primitive recursiveness (Proposition 6). In writing out and then reviewing the definition/proof

$$x|y \Leftrightarrow x \neq 0 \text{ and } \exists n \leq y \, (y = n \cdot x),$$

a prover constructing a schedule of proofs will justify, and thereby come to see the need for explicitly articulating as part of that schedule, the technique of constructing primitive recursive relations by conjunction ('and') and negation ('not') — from which it seeably/showably follows, as a prospective course of proving, that relations constructed through the use of any of the sentential connectives preserve primitive recursiveness in that all the connectives can be defined in terms of 'and' and 'not' alone — and to establish the primitive recursiveness of the equality relation. ($x \neq 0$ can be seen to be defined as not-$(x = 0)$.) In practice, a contemporary prover would most likely not come to retrospectively construct the initial propositions of the schedule in quite this fashion, but instead would begin constructing a schedule by preparing a preliminary list of primitive recursive functions, relations and techniques of constructing such functions and relations and/or by bringing together some of the lists of such functions, relations, and techniques that are available in the literature, selecting from such lists those functions, relations, and techniques found to be needed over the course of constructing the remainder of the schedule and those of practical necessity for that list's orderly development as a coherent course of work.

As a second point, let us note that the justification that $x \neq 0$ defines a primitive recursive relation is the same as that for the relation defined by $y = n \cdot x$; once the justification for either of them is developed as an extractable course of reasoning, the proof of the other follows as well. Specifically, if $y = z$ defines a primitive recursive relation and $n \cdot x$ is a primitive recursive function, then $y = n \cdot x$ defines a primitive recursive relation in that it results from the substitution of a primitive recursive function for a variable in a primitive recursive relation (Proposition 7). But then, since $x \neq y$ is a primitive recursive relation and $Z(w) = 0$ is a primitive recursive function (by definition), the relation defined by 'substituting' $Z(w)$ for y in $x \neq y$ — that is, $x \neq 0$ — is primitive recursive as well.

Finally, the reader should note that (1) the efficacious choice of variables, and the seeable/showable irrelevance of such a choice, is itself a part of a prover's work in producing a notationally consistent derivation of the formula

$x|y \Leftrightarrow x \neq 0$ and $\exists n \leqslant x \, (y = n \cdot x)$

as, for example, the work of such an arranging is available in the 'choosing' of $x \neq y$ and $Z(w) = 0$ above so as to 'deduce' the notationally-specific formula $x \neq 0$, and (2) the fact that a prover can read symbols like 0 as the function $Z(w) = 0$ in order to do what he needs to do with them is tied to the fact that such formulas are already embedded in an ongoing course of work and, further, to a prover's ability, on the occasion of such a need, to historicize that work so as to find, therein, the materially justifiable motives for just such an interpretation and manipulation of symbols.

83 One final digression may help to set in relief the type of constraints that are placed on such work.

Roughly speaking, a function is said to be *effectively computable* if there is a finite, 'mechanical' procedure for calculating the value of that function for any of its possible arguments. A relation is said to be effectively computable if its characteristic function is. Now, in that the initial functions

$Z(x) = 0$

$S(x) = x + 1$

$I_i^m (x_1, \ldots, x_i, \ldots, x_m) = x_i$

can be seen to be computable, and in that the definition of a function by substitution or primitive recursion can be seen to give an effective procedure for calculating the values of that function from the values of the functions used in its definition, the class of functions generated from the initial functions by substitution and primitive recursion — i.e., the primitive recursive functions — would seem therefore to all be computable as well. And, in fact, the definition of the class of primitive recursive functions can be thought of as an (early or preliminary) attempt to formalize the notion of effective computability itself.

Let us consider the relation R consisting of all 3-tuples (y, n, x) such that $y = n \cdot x$. In that, given any 3-tuple of natural numbers, one need only calculate $n \cdot x$ and compare that number to y to see if that 3-tuple is a member of R, it is certainly to be hoped, however the notion of computability is formalized, that R will be able to be shown to be such a relation. In a similar fashion, the divisibility relation can be seen to be 'computable': it is defined by the formula

$x|y \Leftrightarrow x \neq 0$ and $\exists n \, (y = n \cdot x)$

and, as I indicated earlier, y is itself an upper bound on the number n. Thus $3|6$ since $3 \neq 0$ and

$6 \neq 0 \cdot 3 = 0$

$6 \neq 1 \cdot 3 = 3$

$6 = 2 \cdot 3$

but $3 \nmid 5$ in that, if $3 \mid 5$, there would be some number n less than 5 such that $5 = n \cdot 3$ and

$5 \neq 0 \cdot 3 = 0$

$5 \neq 1 \cdot 3 = 3$

$5 \neq 2 \cdot 3 = 6$

$5 \neq 3 \cdot 3 = 9$

$5 \neq 4 \cdot 3 = 12$

$5 \neq 5 \cdot 3 = 15.$

These considerations can be seen to provide heuristic evidence that the divisibility relation and the relation defined by $y = n \cdot x$ are primitive recursive, and, in consequence, they occasion the question of the praxeological origins of their character as being evidential and accountably heuristic, rather than their being corrigibly demonstrative. To answer this question, consider the following: first, various attempts have been made to formalize the notion of (what is referred to as) effective computability; the notion itself has a 'received' history, and from within that history, that notion (and the term 'effective computability') is preserved and maintained as an informal one used in contrast with its various formalizations. Secondly, as part of the received

NOTES TO PAGE 107

history of 'effective computability,' various attempts to formalize it — including the use of Turing computability, Markov algorithms, λ-definability, and general (as opposed to the more restrictive 'primitive') recursion — can all be shown to define the same class of functions; this class of functions properly includes the primitive recursive ones, and, therefore, that a function is 'computable' does not necessarily imply, even heuristically, that it is primitive recursive. (Parenthetically, Church's thesis is the proposal that this larger class of functions *is* the appropriate formulation and formalization of 'effective computability.') Most importantly, however, the circumstance that occasions the preceding remarks and that leads us back into the text is this: the heuristic character of considering the computability of the relation defined by $y = n \cdot x$ consists of the recognition that there is no immediately accessible way, given the local work practices from within which the problem arises (as described in the text) of turning the reasoning concerning the computability of $y = n \cdot x$ into an accountable course of proving.

84 Previously, K_R was referred to 'as' the characteristic function of a relation R, whereas, in the present situation, I have spoken of $K_=$ as 'standing for' the characteristic function of the equality relation. This change in usage points to a phenomenon in its own right. In order to work through the present problem, a prover comes to rely on properties of the function that $K_=$ names. In doing so, the distinction between the symbol $K_=$ and the function for which it stands becomes a feature of the problem's solution. The first point that needs to be made is that the omnipresent 'naming' found in mathematical argumentation — as, for example, in using $K_=$ to 'name' the characteristic function of the equality relation — is available to the mathematician (1) as a practice and (2) in that it is a practice, as a practical resource for the work of proving. Let us make two further observations: First, the formulas $K_=(x, y)$ and $x = y$, as they have come to be written here, are not related as one being 'derived' from the other, but, as the reader will see, they constitute a pair of mutually elaborating expressions. Second, in showing that the equation $x \neq 0$ defines a primitive recursive relation (as will be done later), a prover re-reads 0 as the constant function $Z(y) = 0$. The point of these observations is that they offer the diversity and richness of mathematicians' notational practice as a real-worldly researchable phenomenon and, therein, provide a contrast between that practice and the rendering of it as, for example, in the specification of a logistic system and, as well, in conventional theories of signs and reference.

85 Even though an equation for $K_=(x, y)$ is available,

$$K_=(x, y) = 1 \dotdiv ((x \dotdiv y) + (y \dotdiv x)),$$

a prover does not use such an equation, nor is such an equation enlightening, in showing that $y = n \cdot x$ defines a primitive

217

recursive relation. Instead, as the reader will see below, a prover will 'simply' use the association of $x = y$ with $K_=(x, y)$ or, more exactly, the association of $y = n \cdot x$ with $K_=(y, n \cdot x)$ as it is developed from the association of $y = z$ with $K_=(y, z)$.

86 'Substitute' is placed in quotation marks as a way of signaling the fact that the term refers, at this point, to a manipulation of symbols. Shortly, however, and then retrospectively as well, that term will take on its character as referring to the construction of a function by the process of 'substitution.'

87 And, therein, a prover's recognition of the practically accountable irrelevance of such a choice; therein, once again, a prover's recourse to the material detail of his work as being embedded in the developing course of the work itself.

88 In the display on page 108, $y = z$ is written — or, at least, recognized as the appropriate equation — before the prover writes $K_=(y, z)$; it is from the recognition that the substitution of $n \cdot x$ for z in $y = z$ yields $y = n \cdot x$ that an appropriate choice of variables for $K_=$ is obtained.

89 It might be argued that such a check and, similarly, that some of the 'steps' used in developing the formula $K_=(y, f(n, x))$ are unnecessary. The point here is not that such work is or is not necessary, but that, first, a prover does such work; second, it is through such work that a prover comes to establish $K_=(y, f(n, x))$ as the practically adequate formula for the problem-at-hand; and third, it is against the background of the achieved adequacy of $K_=(y, f(n, x))$ in solving that problem that the essential or incidental character of the work of its production is then assessed.

90 I.e., $K_=(x, y) = 1 \dotminus ((x \dotminus y) + (y \dotminus x))$.

91 Parenthetically, however, a prover might review and/or articulate as a separate proposition some of the techniques of manipulating variables in primitive recursive functions as, for example, the ability to hold all but one of the variables constant when defining a function by substitution.

92 In writing

$$K_T(y, n, x) = K_E(y, f(n, x))$$

after and beneath

$$T(y, n, x) \Leftrightarrow E(y, f(n, x)),$$

a prover will recognize that the former does not necessarily 'follow' from the latter. I will return to this matter later.

93 As the function defined by $K_T(y, n, x) = K_E(y, f(n, x))$ and only incidentally as the characteristic function of the relation defined by $y = n \cdot x$.

94 I.e., '(y, n, x) is a member of T if and only if $(y, f(n, x))$ is a member of E.'

95 For completeness, the following formulation of this proposition and its proof are themselves locally worked out both in relation to each other and as a recurring and circumstantially renewed

NOTES TO PAGE 113

topic over the course of developing a schedule of proofs as a completed whole. Thus, the proposition and proof that follow in the text have their own histories in coming to be the practically adequate statement of that proposition and that statement's practically adequate proof. It should also be noted that the generalizability of both the problem and its solution is already provided for, not because a practically accountable solution has been achieved, but in that the way in which that solution was obtained already incorporated, over the course of its development, the character of its accountable detail as instances of more general and previously recognized constructions.

96 Here, the prover leaves it to the reviewer to find how to establish the fact that the one line follows from the other. For convenience — and as an accountable procedure that promises the generality of its prospective accomplishment — the reviewer might let R be a 2-place relation and f a 1-place function. Then, as in the text above, a method of showing that $S(x, y) \Leftrightarrow R(x, f(y))$ 'implies' $K_S(x, y) = K_R(x, f(y))$ is tied to seeing that $S(x, y)$ can be rewritten in set-theoretic notation as $(x, y) \in S$. As in the text,

$$S(x, y) \Leftrightarrow R(x, f(y))$$

can be written as

$$(x, y) \in S \Leftrightarrow (x, f(y)) \in R$$

and, in that the characteristic function of a subset U of a set V is 1 on U and 0 on V − U, $K_S(x, y)$ can be seen to equal 1 if and only if $K_R(x, f(y))$ equals 1 and $K_S(x, y)$ can be seen to equal 0 if and only if $K_R(x, f(y))$ equals 0.

Two things should be noted: first, that the prover does not elaborate this argument is not because he does not see that such an argument needs to be given, but that in order to make such an argument precise, a larger analytic apparatus would have to be developed, and such work would distract from the larger structure of the accountable work of producing the schedule of proofs, and that, at this point in the prover's work, it is available to the prover that a reviewer should be able to give such an exegesis and to see, in doing so, that that exegesis is incidental to, and does not provide critical detail for, the development of the schedule itself. Second, the reader should also note that while words like 'then,' 'it follows,' 'thus,' and 'implies' are rendered in classical studies of mathematical practice as the notion of 'material implication' − i.e., approximately, from the purely formal expressions $A \supset B$ and A, the prover can then write B — or, possibly, that of 'logical consequence' − i.e., approximately, for every possible assignment of truth values true and false to some set of propositions S, if whenever all $\phi \in S$ have the truth value true, a proposition θ has the truth value true, then θ is a 'logical consequence' of S − in actual mathematical practice these words are used both as indicators and as summaries of

219

situationally-specific, endogenously organized, naturally accountable courses of argumentation. Words like 'then' and 'it follows' serve to articulate the argument being given — for example, by pointing to work that actually needs to be done while, at the same time, maintaining the proof-specific level of appropriate detail, or by calling the readers' attention to what has just been argued in such a way that he thereby comes to see that he needs to reorganize that argument so as to provide for and find the self-evident character of the thing being offered as that argument's consequence — and, in this way, these words take on an endless diversity of what could be called argument-specific meanings.

97 Proposition 6 of the schedule outlined earlier reads 'the logical operations of "not," "and," "or," "implies," and "if and only if," applied to primitive recursive relations, produce primitive recursive relations.' In working through the schedule, this proposition is 'understood' and applied as formulating the fact that if $P(x_1, \ldots, x_m)$ and $Q(x_1, \ldots, x_m)$ are conditions that define primitive recursive relations P and Q, then the expression $P(x_1, \ldots, x_m) \eta Q(x_1, \ldots, x_m)$ defines a primitive recursive relation $P \eta Q$ as well, where η is a sentenial connective and $P \eta Q$ is the relation defined by the appropriate set-theoretic construction on the sets P and Q corresponding to the connective η. The point here is that the cogency and 'meaning' of the statement of this proposition is integrally tied to, and is available to and recognized by the prover as being tied to, the familiar practices of constructing primitive recursive relations that that statement is seen to formulate in an adequate and efficacious manner.

98 I.e., as a retrospectively reconstructed proper order of work.

99 In the fashion of what we have come to speak of as a 'derivation,' this argument can be elaborated as

$x = y$ (5)

$x \neq y$ (6)

$Z(w) = 0$ (by definition)

$x \neq 0$ (7)

100 The reasoning of which can be elaborated, as I did earlier, as follows: in that the equality relation, $y = z$, and multiplication, $n \cdot x$, are, respectively, a primitive recursive relation (by Proposition 5) and a primitive recursive function (by Proposition 3), the equation $y = n \cdot x$ defines a primitive recursive relation in that it results from the 'substitution' of a primitive recursive function for a variable in a primitive recursive relation (Proposition 7); from the primitive recursiveness of $y = n \cdot x$, the technique of bounded quantification (Proposition 8) then insures that $\exists n \leq w \, (y = n \cdot x)$ defines a primitive recursive relation of x, y and w; from which it follows, by 'identifying'

NOTES TO PAGES 115-21

the variables y and w (Proposition 2), that $\exists n \leq y\ (y = n \cdot x)$ defines a primitive recursive relation of x and y alone.

101 At least in principle, and in that they are retrospectively available as having been, and in that they are repaired over the course of developing the schedule as having properly been.

102 *From within* the way of working that is described in topic (d) and *as* an accountable aspect of both the schedule and the syntax that that way of working makes apparent.

103 In the formation rules of a logistic system, the class of terms are typically defined by the following rules: 1) the constant symbols (e.g., 0) are terms, 2) the individual variables are terms, 3) if f is an n-place function symbol and $t_1, \ldots, t_n$ are terms, then $f(t_1, \ldots, t_n)$ is a term, and 4) a formula is a term if and only if it can be shown to be one on the basis of 1), 2) and 3). This definition is to be contrasted to the reformulation of it in terms of a formation sequence of terms as that reformulation is given later in the text.

104 I.e., the (unique) number of terms which, when that number of terms is concatenated with that function symbol and appropriate parentheses, makes up a term itself.

105 Or, depending on the specification of the logisitic system, as a class of expressions.

106 The reader will recall that the individual variables in our system P are x, x', x'', etc. and that the Gödel number that has been assigned to the symbol $'$ is 7.

107 It should be emphasized that the point being made here does not refer to the particular arrangement of these propositions in isolation from the rest of a schedule of proofs, but, instead, concerns the structure of a schedule of proofs as a totality whose natural accountability as a course of proving just that schedule is maintained over that self-same course of proving. The point is that enough of the sequentialized, hierarchical structure of the arithmetized syntax must be present in a schedule of proofs so as to provide access to and to exhibit, in and as that schedule's own material detail, that schedule/the-accountable-work-of-its-production as just such a naturally accountable schedule of proofs. Thus, for the proof of the primitive recursiveness of var(x), any of the three alternative proofs that are given in the text could, in fact, be included in a schedule of proofs, the selection of any one of them not being critical to the exhibited structure of that schedule. This holds similarly for the example of term(x) given below, and in the case of sub(x, n, a), the misplacement of that function's definition/proof in the diagonalization argument would, most likely, appear as an awkwardness to a prover reviewing that proof and, in once again working through the other propositions and proofs of the schedule to find the interpretable 'meaning' of sub(x, n, a) as that 'meaning' is required for the diagonalization/'proof', that prover would, in that way, find as well the proper place for sub(x, n, a) in that schedule.

221

108 One feature of this proof should be noted. The reader will recall from the discussion of the formula

$$x * y = x \cdot \prod_{i=0}^{L(y)} p_{L(x)+i}^{(y)_i}$$

in topic (a) that the term (i.e., the component) $p_{L(x)+0}^{(y)_0}$ (when $i = 0$) was included in the product $\prod_{i=0}^{L(y)} p_{L(x)+i}^{(y)_i}$ specifically to make that formula applicable to the case when $y = 0$. Having discovered this solution to the problem of defining $x * y$, a prover, on confronting the need to make the equation for $g(x_n)$ applicable to the case $n = 0$ as well as that of $n > 0$, can now use that previous solution as, and now has it available as, a device of proving. On coming to and in writing the proof given here, a prover will see the efficaciousness and accountable adequacy of letting the index i include the case $i = 0$ in the product

$$g((x) * \prod_{i=0}^{n} p_i^7 * g(\;))$$

and will, on seeing the utility of that device, know it as just such a previously considered, familiar, accountable procedure.

109 In anticipation of topic (d), we should note that it is the exhibited correspondence between the terms of P and the Gödel numbers of the terms of P that allows a prover to extend the proof of Gödel's theorem for P to logistic systems that properly contain it. (Parenthetically, this is not to say that a prover could not (with additional effort) recover from the preceding formula for term(x) the necessary structure of the Gödel numbering and, therein, the construction of those certain sequences of terms to which the adequacy of that formula is tied. The point, however, which contrasts to this circumstance and which recalls the gestalt theme of the figure/ground is this: a prover, in the course of writing up a proof of a theorem, is engaged in the selection and articulation of those details of the work of that proof that provide for the recovery of that proof as an accountable course of work, and that extractable course of work, in and as a materially-embodied course of argumentation, comes to exhibit those properties of the proof-relevant mathematical objects that are accountably adequate to and, therein, revelatory of the descriptive character of the assertion of the theorem itself. (I am indebted to Harold Garfinkel for this formulation.))

110 The construction that follows bears a recognizable similarity to that of var(x), and, later, to those of wff(x) and ded(x): first, a general property is shown to be primitive recursive — like that of a number x being the Gödel number of a formation sequence of terms — and, then, the desired relation — e.g., term(x) — is shown to consist of those numbers that possess that property and obey some further restriction — in this case,

$$\text{term}(x) \Leftrightarrow \exists y \leq \prod_{i=0}^{L(x)} p_i^x \ (\text{formterm}(y) \text{ and } (y)_{L(y)} = x).$$

Although this type of construction, as a repeating pattern within a schedule of proofs, can be retrospectively disengaged from the schedule as an incidental and nonessential feature of that schedule in providing for the proof of Gödel's theorem, over the course of working through a schedule this repeating construction is part of the exhibited and developing naturalness of the arithmetized syntax as being constructed in programmatic and sequential correspondence to the specification of the original syntax of the system; it is the achievement and maintenance of just this correspondence that makes up the accountable adequacy of the schedule of proofs, and it is this that a prover relies on later when he uses the functions and relations of the arithmetized syntax in the diagonalization/'proof'.

111 This observation has already been illustrated in the discussion of Gödel numbering as a technique of proving. The fact that the notions of a bound and of a free occurrence of a variable in a formula, and that the notion of the substitution of a term for a variable in a wff, are developed in the schedule of proofs in which we are working as the extended series of Propositions 23 through 29 — introducing, in the following order, the functions/relations occur(w, x, y, z), bound(w, x, y, z), free(w, x, y, z), $S^1(x, t, a)$, $S^m(x, t, a)$, $M(x, t, a)$, and, finally, Sub(x, t, a) — is a reminder of the existence of such work and provides a perspicuous place in a schedule for that work's further elucidation.

112 It should be noted that Gödel did not use the logistic system of Whitehead and Russell's *Principia Mathematica* in his original paper but, instead, defined a formal system that, on one hand, was more conducive to his methods of proving his theorem and, therein, to the exhibitable and exhibited adequacy of those methods for such proving and, on the other hand, a system that retained what the new system simultaneously offered as the essentially relevant features of Whitehead and Russell's system for the problem at hand.

113 Parenthetically, the mathematical procedure of using a given materially-specific proof and the structure of proving that that proof both makes available and comes to exhibit to find, through modifications (or variations) of its arguments, the essential or essentially necessary (or ideal) structure of that proof bears a similarity with one interpretation of Husserl's notion of the phenomenological reduction as the free, imagined variation of a perceived object so as to elucidate the essence of that object as an intentional structure of consciousness. Thus, we are here reminded of the possible origins of Husserl's phenomenology in his early training and apprenticeship as a mathematician.

114 The summary characterization that follows stands here in place of a detailed, descriptive review, in the fashion of topics (a) and (b),

115   of the lived-work of developing this part of a schedule of proofs. Here, 'wished' is being used as a gloss for the local and material motives for a prover's inquiry into the accountable structure of a proof of Gödel's theorem as that inquiry is formulated as the need to generalize that proof. Later, in 'A Structure of Proving,' I will indicate how the need to so extend the proof of Gödel's theorem is 'natural' in the sense that it is tied to the accountability of the work of proving Gödel's theorem and, therein, to the structure of the proof of Gödel's theorem itself.

116   I.e., to show that Gödel's theorem holds for such extensions of P as well as for P itself.

117   The appropriateness of such a number is tied to the necessary things that need to be done with the Gödel numbering as those requirements become available over the course of constructing the original proof of Gödel's theorem. Thus, for example, the Gödel number of E could be defined as the first positive odd number greater than those numbers assigned to the primitive symbols of P. The constraints on such a definition, however, do not have to do with the fact that the assigned number is, in and of itself, distinct and/or odd, but, for example, with the fact that if the primitive symbols of the language are all assigned odd numbers, then the Gödel numbers of the expressions and sequences of expressions of the language will all be even numbers (given our proof-specific Gödel numbering), thereby permitting, given any natural number, a strictly mechanical determination of whether or not that number is the Gödel number of a primitive symbol of the language and, if it is, then which one. The notion of the appropriateness of the assignment of Gödel numbers becomes more critical when the language of the formal system is extended to having an infinite number of primitive symbols. This matter will be addressed in the text below.

118   In that the natural numbers are used in the defintiion of the f_r^m, and in that the system P is itself being used to articulate and define (what is then understood as at least part of) the essential structure of the natural numbers, it is not technically correct to introduce the f_r^m as new primitive symbols of P. Although this problem can be circumvented, as in the case of the individual variables, by using the symbol $'$ as in $f_1^2 := ''f'$, the use of such an austere notational system makes the definitions/proofs of the schedule of proofs more difficult and, thereby, obscures the structure of their exhibited reasoning. In that the replacement of the symbols f_r^m with technically correct notation is seen to present no difficulties in principle — i.e., to present no problems other than the technicalities of manipulating the notion for reworking the definitions/proofs of the schedule — the use of f_r^m does not exhibit a damaging fault of a schedule of proofs. As is customary, I have simply used the notational abbreviations f_r^m later in the text as if they were primitive symbols; in particular, I have followed Mendelson's presentation in

NOTES TO PAGES 129-40

Introduction to Mathematical Logic.

119 As I have already indicated, the solution that will be given is found in Mendelson's *Introduction to Mathematical Logic*. Of necessity, some of the details of Mendelson's presentation have been changed to fit the proof of Gödel's theorem as that proof has been developed here.

120 In addition to Mendelson's treatment (pp. 129-30) of course-of-values recursion, the reader is referred to the exposition in Stephan Cole Kleene, *Introduction to Metamathematics* (New York: D. Van Nostrand Company, 1952), pp. 231-3.

121 Again, the solution that is offered here has been taken from Mendelson, *Introduction to Mathematical Logic*.

122 The material in this section does not provide a finished argument, but only some working notes directed to the recovery of a proof's notation as that notation is tied to a prover's work at the mathematical work site. I hope to address this relationship, with materials other than those from a proof of Gödel's theorem, in a separate and later paper.

123 At other places in the schedule, x is used simply to denote a numerical variable as in the definition of the 'decoding' function $(x)_i$, where, as a further example, i is understood as indexing the natural numbers in a serial fashion whereas, in the diagonalization/ 'proof', the letter i is used to stand for the Gödel number of a particular formula.

124 Parenthetically, the reader should note that the serial indexing of m_1, m_2, and m_3 is here used as a local, occasioned and unexplicated device for associating the variables with their appropriate places in the symbolism Sub(x, t, a).

125 I take this occasion to point out a curious feature of the temporal writing of $S_t^x A|$ and, therein, to a more general concern of provers in collaborative work and in classroom and public presentations: the writing of $S_t^x A|$ as $S \rightarrow S^x \rightarrow S_t^x \rightarrow S_t^x A \rightarrow S_t^x A|$ does not correspond to the cotemporaneous description/ elaboration of that writing as denoting the substitution of t for x in A. In order to coordinate that writing and description a prover may prefer to read $S_t^x A|$ as 'the replacement of x by t in A.' Alternatively, a prover may preserve the description and write t prior to x, losing, however, the emphasis on the fact that x is the variable that is being substituted for, or a prover could produce $S_t^x A|$ as a gestalt, maintaining both the temporality of the writing and the description of it as the substitution of t for x in A but announcing, perhaps, 'of t for x' only after having written S_t^x and, then, pointing to t and x as they are named. Three points need to be made: first, in writing a proof, a prover will rely on the seeable things that are provided for by the material notation and not on the verbal names or descriptions of that notation; second, while the temporal coordination of blackboard writing and the description of the notation being used is an omnipresent feature of mathematical presentations, the

NOTES TO PAGES 140-7

exact manner in which blackboard displays are produced has not, to my knowledge, been subjected to examination; and third, the nature and consequentiality of that coordinated work, for mathematical practice, cannot be decided prior to an actual examination of those practices.

126 Quine, for example, adopts the convention of indicating the variable x being substituted for as the subscript and the term t being substituted as the superscript, writing, in his notation, $x'\ SF_y^y\ x$ to mean that the formula x' results from the substitution of y' for the free occurrences of y in x. (Willard Van Orman Quine, *Mathematical Logic*, revised edn (Cambridge: Harvard University Press, 1951), p. 301.) By modifying the notation used in the text, this would result in writing $S_x^t\ A|$ and, thereby, of bringing its description as 'the-substitution of the term t for x in A' into accord with the temporality of its writing.

127 Joel W. Robbin, (New York: Benjamin/Cummings Publishing Company, 1969).

128 The reader will note, however, that I did not make such changes uniformly throughout the schedule of proofs for reasons tied, for example, to the interpretability of the notation for Sub(x, t, a).

129 The variables in occur(w, x, y, z) are arranged in alphabetical order, exhibiting, in that way, the arbitrariness of their selection and, therein, the fortuitousness of x corresponding (if occur (w, x, y, z) holds) to the Gödel number of an individual variable.

130 An initial candidate definition for freefor(x, t, a) might be the relation T(t, x, a) defined by

T(t, x, a) ⇔ term(t) and var(x) and wff(a) and

$\exists a_1 \leqslant a\ \exists a_2 \leqslant a$ {free(a_1, x, a_2, a) implies not-$\exists c_1 \leqslant a$

$\exists c_2 \leqslant a\ \exists y \leqslant a\ \exists t_1 \leqslant t\ \exists t_2 \leqslant t$ [var(y) and $L(c_1) \leqslant L(a_1)$

and $L(c_2) \leqslant L(a_2)$ and bound(c_1, y, c_2, a)]} .

T(t, x, a), however, is too restrictive a definition in that the bound occurrence of y in A indicated by bound(c_1, y, c_2, a) may occur in a well-formed part of A not containing the free occurrence of x. The more extended condition given later in the text remedies this problem.

131 Where that arbitrariness, and the opacity of that notation, is now to be understood in terms of the work associated with a schedule of proofs' notational development.

132 K. Koffka, *Principles of Gestalt Psychology* (New York: Harcourt, Brace and Co., 1935), p. 172.

133 Although, as we have seen in point 1 of this chapter, a prover may use the specific assignment of Gödel numbers to develop some of the proofs of a schedule, the fact that that work has no residue in the finished schedule does not mean that the prover is concealing that work from his reader. A prover is held to exhibit, in and as the material detail of a proof, the work of proving that is

adequate to, and exhibiting of, that material detail as composing a naturally accountable proof. No residue of the use of specific Gödel numbers is found in the schedule in that that work is not work that is tied to the naturally accountable schedule of proofs. It is the lack of that connection which gives to a prover's illustration of and commentary on such work its character as only heuristic and pedagogically directed remarks.

134 Generally, in the way in which the Gödel numbering is otherwise being used; specifically, as the reader will see shortly, in a manner similar to that discussed later in the text.

135 As I did in some of the notes of point 1 of this chapter.

136 Such notation probably being introduced at the same time that the Gödel numbering is defined.

137 J. Donald Monk adopts this notation in his *Mathematical Logic* (New York: Springer-Verlag Inc., 1976).

138 It should be pointed out that, given the need for different uses of a Gödel numbering, the designation of functions g, g^+, and g^{++}, and designating them in such a way so as to suggest the relationship between them, may be critical to the development of some mathematical arguments. The reader is again referred to J. Donald Monk, *Mathematical Logic* (New York: Springer-Verlag Inc., 1976).

Chapter 8 A Structure of Proving

1 I.e., the rule of inference that states from 'A' and 'A implies B', one can infer B.

2 Cf. the proof in Chapter 2 of the uniqueness of the identity element of a group.

3 L. Chwistek gives an abbreviated proof, in a particular formal system, of a theorem that he describes as 'correspond[ing] to the first theorem of Gödel,' but where 'the contradiction does not appear in $\vdash_0 0$ but in $\vdash_2 2$.' Chwistek states this theorem as

$$\vDash [10] \supset \vee \vdash_0 G_{(2)} \vdash_0 \sim G_{(2)} \vdash_0 \vdash_2 = .0(5).1(5)[5] \text{ is a theorem}$$

where $G_L(E)$ is an abbreviation of

$$\exists [.0(L)L] \; \bar{x}_L \bar{y}_L \bar{z}_L \bar{u}_L \wedge = E \; \Pi \; [.2(L).1(L)] \bar{y}_L \bar{z}_L [.0(L)]$$
$$\wedge \mathrm{Im}_{.0(L)} E\bar{x}_L \wedge (\bar{z}_L \bar{y}_L \bar{x}_L \bar{u}_L) [.0(L)] \sim \vdash_L \bar{u}_L$$

and $G_{(L)}$ is an abbreviation of

$$G_L(\Pi \; [.2(L).1(L)] \; \bar{a}_{.1(L)} \; G_{.1(L)} \; (\bar{a}_{.1(L)})).$$

This theorem allows him to prove 'Gödel's [second] theorem'

$$\vDash [10] \supset \vdash_0 (32) \sim \vdash_2 (54) = .0(5).1(5)[5] \vdash_0 (32) = .0(3).1(3)[3]$$

in a manner that 'is quite formal and does not differ in any way

NOTES TO PAGES 150-4

from ordinary arithmetical proofs.' The relevant articles here are L. Chwistek and W. Hetper, 'New Foundations of Formal Metamathematics,' *Journal of Symbolic Logic*, 3, 1, March 1938, pp. 1-36 and L. Chwistek, 'A Formal Proof of Gödel's Theorem,' *Journal of Symbolic Logic*, 4, 2, June 1939, pp. 61-9.

4 I am indebted to Harold Garfinkel for this last observation.

5 The notion of an organizational object that I am using here is due to Harold Garfinkel, and I have relied on his writings and lectures in developing the list that follows. Responsibility for the material in the text is my own.

6 'Anyone' refers to the fact that the proof is available to provers as a practically objective thing, as an 'organizational object.' For provers, if someone cannot see that such a proof has been given, if a person cannot see that the work done to prove the theorem was the work that was required of that proof, then reasons are given for that person's inability to 'understand' that proof – e.g., that that person does not have a 'mathematical mind,' that he has not been 'trained' as a mathematical logician, etc. (Parenthetically, in that it is the inability of someone to do the work that makes up the recognizably adequate proofs of mathematics that exhibits that person's mathematical incompetence (and, conversely, in that it is the ability of someone to do that work that identifies him, in the company of other provers, as a prover himself), what the formulations of a person's mathematical incompetence can be seen to always lack is the specification of the features of the lived-work of proving, from within that lived-work, that are identifying of that work as such.)

7 E.g., on seeing the proof that a homomorphism ϕ of a group A to a group B maps the identity 1_A of A to the identity 1_B of B – the main 'idea' of which is that, by writing 1_A as $1_A\, 1_A$, one can write

$$\phi(1_A) = \phi(1_A 1_A) = \phi(1_A)\, \phi(1_A)$$

– a beginning prover may not see the 'relevance' of this same 'idea' to proving that $\phi(x^{-1}) = [\phi(x)]^{-1}$ thus,

$$1_B = \phi(1_A) = \phi(xx^{-1}) = \phi(x)\, \phi(x^{-1}).$$

8 As the reader will see, the expression 'a structure of proving' is intended as an initial, descriptive formulation of the phenomenon of which it speaks. Thus, the problems that remain are not those of analyzing and constructively interpreting the notion of a 'structure of proving,' but those of using that notion, if possible, to gain deeper and more technical access to mathematicians' work practices. By pointing out that the notion of a 'structure of proving' provides only a provisional solution to the characterization problem, I wish to emphasize that it remains for later investigations to come to a more competent understanding of the characterization problem.

9 As I have already indicated, a more precise statement of the theorem is that, for the particular sentence J constructed in the

diagonalization argument, (1) if P is consistent, then *J* is not deducible in P, and (2) if P is ω-consistent, then $\sim J$ is not deducible in P.

10 I.e., for every wff of P with no free variables.

11 Parenthetically, the reason Gödel's 'second' theorem is also referred to as an incompleteness result is that that theorem 'says' that one of the 'undecidable' sentences of formal number theory is the (a) sentence that 'arithmetically' asserts that that theory is consistent; thus, the more common formulation — if the theory of arithmetic is consistent, then its consistency cannot be demonstrated within the theory itself.

12 Moreover, in that *J* can be interpreted as 'saying' that it itself is not deducible in P, and in that *J* is not deducible in P by the proof of Gödel's theorem given earlier, *J* represents an 'intuitively' true proposition of elementary arithmetic (where 'intuitively' refers to the informal character of the procedure whereby *J* is given its interpreted meaning).

13 In that the non-logical axioms of P are true propositions of number theory under their standard interpretation in N, and in that the logical axioms and rule of inference of P correspond as abstractions or epitomizing characterizations of familiar and regularly recurring techniques of naturally accountable mathematical reasoning, the difficulty in not being able to show that P is complete — assuming that ordinary number theory is itself consistent and complete — does not seem to reside in the fact that P is too strong a theory in the sense that restrictions in the specification of P might result in a system P^- that still 'represented' number theory but in which a comparable statement of Gödel's theorem did not hold. This reasoning underlies the discussion of the possible extension of P to a complete theory. Two comments, however, are appropriate. First, one of the attempts to circumvent the apparent achievement of Gödel's second incompleteness theorem in regard to the formal investigation of elementary arithmetic was based on the diagnosis that the proofs of Gödel's theorems depended on an unrestricted use of universal quantification and that, in this way, P is too strong a theory. (See Alonzo Church, *Mathematical Logic*, Lectures at Princeton University, October 1935–January 1936, notes by F.A. Ficken, *et al.*) Second, Gerhard Gentzen was able to prove the consistency of a system representing elementary arithmetic, but used the inferential procedure of transfinite induction up to the ordinal

$$\epsilon_0 = \omega \omega^{\omega^{\cdot^{\cdot^{\cdot}}}}$$

in order to do so. Transfinite induction up to ϵ_0 is not itself deducible in our system P, and, because of this, Gentzen's proof of the consistency of 'arithmetic' may be said to rely on methods stronger than those used in 'arithmetic' itself.

Gentzen's original papers appear in translation in M.E. Szabo (ed.), *The Collected Papers of Gerhard Gentzen* (Amsterdam: North-Holland Publishing Company, 1969). The reader is also referred to discussions of Gentzen's work in Elliot Mendelson, *Introduction to Mathematical Logic* (Princeton: D. Van Nostrand Company, Inc., 1964), Howard Delong, *A Profile of Mathematical Logic* (Reading: Addison-Wesley, 1970), Stephan Cole Kleene, *Introduction to Metamathematics* (New York: D. Van Nostrand Company, 1952), and Gaisi Takeuti, *Proof Theory* (Amsterdam: North-Holland Publishing Company, 1975).

14 The reader may wish to argue that, properly understood, the proof of Gödel's theorem, as it has been presented for P, already answers this question. What is at stake, however, is not the fact of whether this is or is not so, but the specification of the work required on the part of a prover to see that this is so.

15 The syntax of this system will not be precisely defined, and the reader is referred to conventional sources for such definitions. Particularly careful treatment of such matters is to be found in Alonzo Church, *Introduction to Mathematical Logic* (Princeton: Princeton University Press, 1956), Vol. 1. For the material that follows, I have used, and wish to acknowledge my indebtedness to, Elliot Mendelson, *Introduction to Mathematical Logic* (Princeton: D. Van Nostrand, 1964) and Joel W. Robbin, *Mathematical Logic: A First Course* (New York: Benjamin/Cummings Publishing Company, 1969).

16 I will not make explicit all the parentheses in the formulae that follow, nor will I place the function symbols $+$ and $\cdot$ to the left of their arguments as would be required by an explicit statement of the formation rules of the system. In stating the axioms, such conformity to the formal specification of the syntax of P would only obfuscate the meaning of the axioms as interpreted in N. On the other hand, it is important to note the presence of such abbreviations in that, in later constructing the schedule of proofs, we will use the unabbreviated form of some of these axioms.

17 In this note, I indicate a technical revision of the specification of the axioms of P.

A first-order theory T is said to be a first-order theory with equality if $x_1 = x_1$ and $x = y \supset (A \supset A')$ are theorems of T, where x and y are individual variables, A is a wff, A' is obtained from A by replacing some, but not necessarily all, occurrences of x in A by y' and y is free for x at those places in A where it replaces x. (Eliott Mendelson, *Introduction to Mathematical Logic* (Princeton: D. Van Nostrand, 1964), p. 75). Define the atomic formulas of a theory T to be the formulas of the form $A_i^n(t_1, \ldots, t_n)$ where A_i^n is an n-place predicate symbol of T and $t_1, \ldots, t_n$ are terms of T. (In P, the only predicate symbol is $=$, and the atomic formulas are written in the form $t_1 = t_2$.) Then, it can be shown that if $x_1 = x_1$ and if $x = y \supset (A \supset A')$

hold for the atomic formulas A of a first-order theory T, then T is a theory with equality. (Mendelson, p. 76.) Let us number these last two formulas $6'$ and $7'$, respectively. Now, while axioms 6 and 7 in the text are themselves sufficient to show that P is a first-order theory with equality (Mendelson, p. 107), it is customary and convenient to distinguish between a basic underlying logistic system (of which the quality relation, if present, is considered a part) and the specific axioms of a theory under construction. The axioms of P can be brought into accord with this policy by replacing axioms 6 and 7 with new axioms $6'$ and $7'$, and the new axiom $7'$ can be introduced into the schedule of proofs as follows: First, define atomic(a) as holding if and only if a is the Gödel number of an atomic formula of P. atomic(a) is primitive recursive since

atomic(a) $\Leftrightarrow \exists t_1 \leq a\; \exists t_2 \leq a$ (term(t_1) and term(t_2) and

$a = t_1 * g(=) * t_2$).

Then

axiom$_{7'}$(w) $\Leftrightarrow \exists x \leq w\; \exists y \leq w\; \exists a \leq w\; \exists b \leq w$ {var(x) and

var(y) and atomic(a) and L(a) = L(b) and

$\forall i \leq L(a)\; ((a)_i = (b)_i$ or $[(a)_i = x$ and $(b)_i = y])$

and $w = g((() * x * g(=) * y * g() \supset () * a * g(\supset) *$

$b * g()))$ } .

Since a is the Gödel number of an atomic formula A — from which it follows that there are no quantifiers in A — we need not be concerned over whether or not y is free for x in A. (The idea of this definition/proof was taken from Herbert B. Enderton, *A Mathematical Introduction to Logic* (New York: Academic Press, 1972), p. 224.)

18 I have made no attempt at originality in developing the 'schedule of proofs' that follows; most of the statements and proofs of it follow closely those found in Joel W. Robbin, *Mathematical Logic: A First Course* (New York: Benjamin/Cummings Publishing Company, 1969). I have made some changes — in wording and detail — and, as always, responsibility for the text is my own. In addition to these minor changes, I have modified Robbin's presentation in two other ways: First, some modifications were necessitated by the fact that Robbin proves Gödel's theorem for first-order primitive recursive arithmetic, whereas I provide a proof of it for the system P. Second, Robbin, as is now customary as well as pedagogically efficient (both for reasons internal to the proof of Gödel's theorem and in order to introduce topics — e.g., primitive recursive functions and relations — independently of their use in a proof of that theorem), divides the schedule of proofs into distinct, topically identified sections. In this, I have instead followed Gödel's original

formating of the propositions as a displayed 'schedule.' On the other hand, however, Gödel introduces the necessary elementary primitive recursive functions and relations and techniques for constructing primitive recursive functions and relations prior to beginning the schedule itself; his schedule begins with the assertion of Proposition 11 below. (See Kurt Gödel, 'On Formally Undecidable Propositions of *Principia Mathematica* and Related Systems I,' in translation from German in Jean van Heijenoort, *From Frege to Gödel: A Source Book in Mathematical Logic 1879-1931* (Cambridge: Harvard University Press, 1967), pp. 596-616. Finally, in constructing a 'diagonalization/ "proof"' for the proof of Gödel's theorem, I made use of the 'diagonalization/"proof"' found in Raymond Wilder, *The Foundations of Mathematics*, 2nd edn (New York: John Wiley and Sons, 1965), pp. 272-3.

19 In that the construction of a schedule of proofs is informed by the project of arithmetizing the syntax of P, one last proposition offers itself for inclusion in the schedule, namely

36 theorem(a), which holds if and only if a is the Gödel number of a theorem, is a primitive recursive relation,

as well as a potential proof of it,

theorem(a) $\Leftrightarrow \exists \lambda \leqslant f(a) \{ded(\lambda, a)\}$,

where f(a) is a primitive recursive function of a. However, the construction of such a function f(a) is not forthcoming; in fact, if such a primitive recursive function existed — and, hence, if theorem(a) was a primitive recursive relation — then, given any formula M of P, one could compute $g(M)$ and, in principle, calculate whether or not M was deducible in P. This, however, contradicts a consequence of Church's theorem that if P is ω-consistent, then there can be no 'effective procedure' for determining, for an arbitary wff M, if M is or is not a theorem of P. Explicit mention of the relation theorem(a) is made, for example, in the schedules of proofs found in Gödel's original paper, in Robbin, *Mathematical Logic: A First Course*, and in Joseph R. Shoenfield, *Mathematical Logic* (Reading: Addison-Wesley, 1967).

20 A proof of this theorem can be found in Elliot Mendelson, *Introduction to Mathematical Logic* (Princeton: D. Van Nostrand, 1964), pp. 131-4.

21 For example, see the proof of Elliot Mendelson, *Introduction to Mathematical Logic* (Princeton: D. Van Nostrand, 1964), pp. 131-4.

22 Throughout this book, I have used the notion of P being a 'model' of N in the ordinary sense of P being a representation of it. In mathematical logic, however, a model $\mathfrak{M}$ of a set of sentences is defined by first defining the notion of logical implication (indicated notationally as $\models_{\mathfrak{M}} \sigma$) as distinguished from

NOTES TO PAGES 167-8

the notion of deducibility (with its notational designation $\vdash_P \sigma$). Given this definition, a sentence σ is *true* in $\mathfrak{M}$ (and $\mathfrak{M}$ is said to be a model of σ) if and only if $\vDash_{\mathfrak{M}} \sigma$. is said to be a model of a set Σ of sentences if and only if it is a model of every member of Σ. (See Herbert B. Enderton, *A Mathematical Introduction to Logic* (New York: Academic Press, 1972), p. 84 and the pages immediately preceding p. 84 for preparatory material.) In this sense, then, $N = \{|N|, 0, S, +, \cdot\}$ is said to be a model of the axioms of P, thus reversing our way of informally speaking of P as a 'model' of N.

23 For a little more detail, however, see note 17 on p. 230. Besides the criticisms found there, other criticisms of the reasoning in the text above are possible as, for example, the fact that our system P is not finitely axiomizable. What I wish to stress here, however, is not the endless refinability of Gödel's theorem and its proof, but the fact that the proof of that theorem is witnessibly exhibiting of its naturally accountable accomplishment. I will briefly return to this matter at the beginning of section C of this chapter.

24 As noted later in the text, such an extension need not be limited to the language of P, but may be an extension of that language as well.

25 One of the aims of the next section is to begin to elucidate what it means 'to see' that this is so. Robbin, for example, writes: 'It is, of course, not particularly surprising that a particular set of axioms is not complete. The full impact of the Gödel theorem comes *on realizing* that the proof of the Gödel theorem works for any reasonable set of axioms for arithmetic.' (Joel W. Robbin, *Mathematical Logic: A First Course* (New York: Benjamin/Cummings Publishing Company, 1969), p. 115, italics mine.) The problem that I want to point to is that of determining what, as practice, such a 'realization' consists of. I now quote Robbin's 'concluding' remarks in full as an illustration of how an author might 'temporarily close' the discussion of the proof of Gödel's first theorem. (For my qualification concerning the temporary character of the closure of the discussion of the proof the reader is again referred to the next section of this chapter.)

> It is, of course, not particularly surprising that a particular set of axioms is not complete. The full impact of the Gödel theorem comes on realizing that the proof of the Gödel theorem works for any reasonable set of axioms for arithmetic. So long as the set of Gödel numbers of axioms is primitive recursive, the Gödel technique [the diagonalization procedure] can be applied to yield a true sentence which is not provable. Thus, if we form a new formal language RA' by adding J as an axiom to the axioms of RA [first-order primitive recursive arithmetic], we have a new two-place predicate $\text{ded}'(\sigma, a)$, which says that σ is the Gödel number of a deduction in RA'

of the wff with Gödel number a, and by exactly the same construction as above, we get a sentence J' which is true but is not a theorem of RA'. Thus, as long as we demand of a formal language that there be an effective procedure for recognizing which sequences of wffs are deductions (i.e., $ded(\sigma, a)$ is primitive recursive), there is no formal language whose theorems consist of precisely the true sentences of arithmetic.

26 These axioms have been taken from Herbert B. Enderton, *A Mathematical Introduction to Logic* (New York: Academic Press, 1972), p. 194.
27 See pp. 125-37.
28 Barkley Rosser, 'Extensions of Some Theorems of Gödel and Church,' *Journal of Symbolic Logic*, 1, 3 (1936), pp. 87-91.
29 In his paper 'An Unsolvable Problem of Elementary Number Theory,' Alonzo Church writes in a footnote: 'This definition is closely related to, and was suggested by, a definition of recursive functions which was proposed by Kurt Gödel, in lectures at Princeton, N.J., 1934, and credited by him in part to an unpublished suggestion of Jacques Herbrand. The principal features in which the present definition of recursiveness differs from Gödel's are due to S.C. Kleene,' *American Journal of Mathematics*, Vol. LVIII, no. 2 (1936), p. 351, n. 9.
30 Alonzo Church, 'An Unsolvable Problem of Elementary Number Theory,' *American Journal of Mathematics*, 58, 2 (1936), pp. 345-63.
31 Alfred Tarski, 'The Concept of Truth in Formalized Languages,' trans. from the German translation of the original Polish text (1933) by J.H. Woodger in Alfred Tarski, *Logic, Semantics, and Metamathematics* (Oxford: Clarendon Press, 1956), pp. 152-278. See as well Robbin's treatment in Joel W. Robbin, *Mathematical Logic: A First Course* (New York: Benjamin/Cummings Publishing Company, 1969), pp. 119-21.
32 Gerhard Gentzen, 'Die Widerspruchsfreiheit der reinen Zahlentheorie,' *Mathematische Annalen*, 112 (1936), pp. 493-565, translated as 'The Consistency of Elementary Number Theory' in M.E. Szabo (ed.), *The Collected Papers of Gerhard Gentzen*; 'Neue Fassung des Widerspruchsfreiheitsbeweises für die reine Zahlentheorie,' *Forschungen zur Logik and zur Grundlegung der exakten Wissenschaften*, New Series, no. 4, Leipzig (Hirzel), (1938), pp. 19-44, translated as 'New Version of the Consistency Proof for Elementary Number Theory' in M.E. Szabo (ed.), *op. cit.*, pp. 252-86.
33 It should be noted that it is not necessary for a proof of Gödel's theorem to incorporate *this* self-referential feature explicitly; a more 'contemporary' way of proving Gödel's theorem is to first prove a 'diagonal lemma,' to use that lemma to show that the set of Gödel numbers of the theorems of any consistent extension of P is not recursive (Church's theorem), to show that if a formal

NOTES TO PAGES 171-81

system is axiomatized and complete, the set of Gödel numbers of the theorems of the system is recursive, and, in consequence, to conclude that any consistent extension of P is incomplete. (See, for example, Joseph R. Shoenfield, *Mathematical Logic* (Reading: Addison-Wesley Publishing Company, 1967), pp. 131-2). The relevance of the self-referential character of Gödel's (and our) proof to the discussion of what identifies a proof of a particular theorem as a proof of just that theorem is somewhat indirect and is indicated in the text that immediately follows.

34 I am indebted to Harold Garfinkel for the observation that a proof's achievement is the pointlessness of disputing that proof's reasoning and the properties of the mathematical objects that that proof makes descriptively available. Responsibility for the use of this expression in the text is, of course, my own.

35 The formulation of a proof as consisting of such a pair is due to Harold Garfinkel. He speaks of the first part of the pair not as the 'material proof' but as a 'proof account.' The pointedness of this observation will be briefly indicated later and in the next chapter.

Chapter 9 Summary and Directions for Further Studies

1 The discovery and formulation of 'classical studies of work,' the availability of 'classical studies' as a phenomenon from within ethnomethodological studies of work, and the discovery and formulation of the discovering sciences as classical sciences of practical action are all due to Harold Garfinkel. I am indebted to him for making this material freely available. I am indebted to him as well for the overall perspective from within which this section was written. Responsibility for the text, however, is mine alone.

2 The discovery and formulation of a mathematical proof as a classical study of practical action, and the discovery and formulation of mathematics in particular, and of the discovering sciences in general, as classical sciences of practical action and practical reasoning are due to Harold Garfinkel. I am indebted to him for offering, and thus for bringing my studies to, these both epitomizing and deeply suggestive formulations of mathematicians' work.

Appendix

1 A reference for the material from physics used in this appendix is Richard P. Feynman, Robert B. Leighton and Matthew Sands, *The Feynman Lectures on Physics* (Reading: Addison-Wesley, 1964), Vol. II, Chapters 2 and 3. Michael Spivak's *Calculus on*

Manifolds (Menlo Park: W.A. Benjamin, 1965) is a source for the material from mathematics.

2 This appendix was written before the book. At this point in the appendix, an attempt to abstractly formulate what it means to speak of the relationship between mathematics and physics as a problem in the production of social order would introduce a severe distraction into its presentation. A sense of what this means should become clearer by the end of the appendix. The reader is asked to wait until he has read the body of the book to begin to see how the phrase 'a problem in the production of social order' takes on its descriptive sense within a technical argument, its use being animated by and clarifying of the material details of the mathematicians' praxis.

3 The statement of the theorem and the material concerning it is taken primarily from Spivak, *Calculus on Manifolds*.

4 I speak here of two 'reasons' for the pervasiveness of this presupposition. What is intended, however, is an initial way of locating the embeddedness of conventional analyses of mathematical physics within the methods of doing those analyses.

5 Very briefly, Fourier's law is an imprecise law that says that the flow of heat in an isotropic medium is, infinitesimally, in the direction of the maximum decrease of temperature and in magnitude, proportional to that decrease. Fourier's law can be empirically checked, whereas the adequacy of the heat equation is found by comparing its solution, subject to the specification of initial and boundary conditions, to the empirical object. The heat equation is a linear partial differential equation that gives the local law of the change of temperature. In the discussion that follows, I am not concerned with proper constants (as κ) and write $\Delta u \stackrel{ess}{=} u_t$ for an essential equality. Δ is the Laplacian; the subscript t stands for partial differentiation with respect to time.

6 This appendix, having been written prior to the body of the book, has postponed the analysis of mathematicians' praxis. Nevertheless, I point out here that the methods of reading theoretical-physics-as-mathematics simultaneously interrogate and organize a text so as to elucidate, and therein build, a properly mathematical argument. From within these practices, the mathematics of theoretical physics stands immediately juxtaposed to its potentially mathematical rendering. These practices stand in contrast to the diverse and as of yet unanalyzed ways in which physicists examine such writing from within their praxis as theoretical physicists.

Bibliography

Baccus, Melinda. 'Sociological Indication and the Visibility Criterion of Real World Social Theorizing.' Unpublished paper, Department of Sociology, UCLA, 1976.
Baccus, Melinda. 'Multipiece Truck Wheel Accidents and Their Regulation.' Unpublished paper, Department of Sociology, UCLA, 1981.
Barnes, Donald W. and Mack, John M. *An Algebraic Introduction to Mathematical Logic.* New York: Springer-Verlag Inc., 1975.
Bellman, Beryl L. *The Language of Secrecy.* Unpublished monograph, Department of Sociology, University of California, San Diego, 1981.
Boolos, G. and Jeffrey, R. *Computability and Logic.* Cambridge: Cambridge University Press, 1974.
Burns, Stacy. 'A Comparison of Geertzian and Ethnomethodological Methods for The Analysis of the Production of Member-Relevant Objects.' Unpublished paper, Department of Sociology, UCLA, 1978.
Burns, Stacy. 'A Comparison of March-Olson and Ethnomethodology on "Circumstantial" Analysis of Decision-Making Work.' Unpublished paper, Department of Sociology, UCLA, 1978.
Burns, Stacy. 'Lecturing's Work.' Unpublished paper, Department of Sociology, UCLA, 1978.
Burns, Stacy. 'Becoming a Lawyer at Yale Law School.' Unpublished paper, New Haven: Yale Law School, 1981.
Choquet, Gustave. *Geometry in a Modern Setting.* Boston: Houghton Mifflin Company, 1969.
Church, Alonzo. 'An Unsolvable Problem of Elementary Number Theory.' *American Journal of Mathematics* 58, 2 (1936): 345-63.
Church, Alonzo. *Mathematical Logic.* Lectures at Princeton University, 1935-6. Notes by F.A. Ficken, H.G. Landau, H. Ruja, R.R. Singleton, N.E. Steenrod, H.H. Sweer and F.J. Weyl.
Church, Alonzo. *Introduction to Mathematical Logic*, Vol. I. Princeton: Princeton University Press, 1956.
Chwistek, L. 'A Formal Proof of Gödel's Theorem,' *Journal of Symbolic Logic* 4, 2 (1939): 61-9.

BIBLIOGRAPHY

Chwistek, L. and Hetper, W. 'New Foundations of Formal Mathematics.' *Journal of Symbolic Logic* 3, 1 (1938): 1-36.

Delong, Howard. *A Profile of Mathematical Logic*. Reading: Addison-Wesley Publishing Company, 1970.

Eglin, Trent. 'Introduction to a Hermeneutics of the Occult: Alchemy.' In E. Tiryakian (ed.), *On the Margin of the Visible: Sociology, the Esoteric, and the Occult*. New York: John Wiley and Sons.

Enderton, Herbert B. *A Mathematical Introduction to Logic*. New York: Academic Press, 1972.

Fauman, Richard. 'Filmmakers' Work: On the Production and Analysis of Audio-Visual Documents for Social Sciences.' Unpublished paper, Department of Sociology, UCLA, 1980.

Feferman, F. 'Arithmetization of Metamathematics in a General Setting.' *Fundamenta Mathematicae* 49 (1960): 35-92.

Feynman, Richard P., Leighton, Robert B. and Sands, Matthew. *The Feynman Lectures on Physics*. Vol. II. Reading: Addison-Wesley Publishing Company, 1964.

Garfinkel, Harold. 'A Conception of and Experiments with "Trust" as a Condition of Concerted Stable Actions.' In O.J. Harvey (ed.), *Motivation and Social Interaction*. New York: The Ronald Press Company, 1963.

Garfinkel, Harold. *Studies in Ethnomethodology*. Englewood Cliffs: Prentice-Hall, Inc., 1967.

Garfinkel, Harold and Burns, Stacy. 'Lecturing's Work of Talking Introductory Sociology.' Unpublished paper, Department of Sociology, UCLA, 1979.

Garfinkel, Harold, Lynch, Michael and Livingston, Eric. 'The Work of the Discovering Sciences Construed with Materials from the Optically Discovered Pulsar.' *Philosophy of the Social Sciences* 11 (1981): 131-58.

Garfinkel, Harold and Sacks, Harvey. 'On Formal Structures of Practical Actions.' In John C. McKinney and Edward Tiryakian (eds), *Theoretical Sociology: Perspectives and Developments*. New York: Appleton-Century-Crofts, 1970.

Gentzen, Gerhard. 'Die Widerspruchsfreiheit der reinen Zahlentheorie.' ('The Consistency of Elementary Number Theory.') *Mathematische Annalen* 112 (1936): 493-565. English translation appearing in M.E. Szabo (ed.), *The Collected Papers of Gerhard Gentzen*, pp. 132-213.

Gentzen, Gerhard. 'Neue Fassung des Widerspruchsfreiheitsbeweises für die reine Zahlentheorie.' ('New Version of the Consistency Proof for Elementary Number Theory.') *Forschungen zur Logik und zur Grundlegung der exakten Wissenschaften*, New Series, 4, Leipzig (Hirzel), (1938): 19-44. English translation appearing in M.E. Szabo (ed.), *The Collected Papers of Gerhard Gentzen*, pp. 252-86.

Gödel, Kurt. 'Über formal unentscheidbare Sätze der *Principia Mathematica* und verwandter Systeme I.' ('On Formally Undecidable Propositions of *Principia Mathematica* and Related Systems I.') *Monatschefte für Mathematik und Physik* 38 (1931): 173-98.

English translation appearing in Jean van Heijenoort, *From Frege to Gödel: A Source Book in Mathematical Logic, 1879-1931*, pp. 596-616.
Gurwitsch, Aron. *The Field of Consciousness.* Pittsburgh: Duquesne University Press, 1964.
Kleene, Stephan Cole. *Introduction to Metamathematics.* New York: D. Van Nostrand Company, Inc., 1952.
Koffka, K. *Principles of Gestalt Psychology.* New York: Harcourt, Brace and Co., 1935.
Leiberman, Kenneth. *Understanding Interaction in Central Australia: An Ethnomethodological Study of Australian Aboriginal People.* Doctoral Dissertation, Department of Sociology, University of California, San Diego, 1980.
Leiberman, Kenneth. 'The Economy of Central Australian Aboriginal Expression: An Inspection from the Vantage of Merleau-Ponty and Derrida.' *Semiotica*, in press.
Levi, Howard. *Foundations of Geometry and Trigonometry.* Englewood Cliffs: Prentice-Hall, Inc., 1960.
Livingston, Eric. 'An Ethnomethodological Approach to the Arts.' Unpublished paper, Department of Sociology, UCLA, 1976.
Livingston, Eric. 'Mathematicians' Work.' Paper presented in the session 'Ethnomethodology: Studies of Work,' *Ninth World Congress of Sociology*, Uppsala, Sweden, 1978.
Lynch, Michael. *Art and Artifact in Laboratory Science: A Study of Shop Work and Shop Talk in a Research Laboratory.* Doctoral Dissertation, School of Social Sciences, University of California, Irvine, 1979.
Lynch, Michael. *Art and Artifact in Laboratory Science.* London: Routledge & Kegan Paul, 1985.
Lynch, Michael, Livingston, Eric and Garfinkel, Harold. 'Temporal Order in Laboratory Work.' In Karen Knorr and Michael Mulky (eds), *Science Observed: Perspectives on the Social Study of Science.* London; Sage Publications Ltd., 1983.
Lyndon, Roger C. *Notes on Logic.* Princeton: D. Van Nostrand Company, Inc., 1966.
Mendelson, Elliott. *Introduction to Mathematical Logic.* Princeton: D. Van Nostrand Company Inc., 1964.
Merleau-Ponty, Maurice. *Phenomenology of Perception.* Translated by Colin Smith. London: Routledge & Kegan Paul, 1962.
Monk, Donald J. *Mathematical Logic.* New York: Springer-Verlag Inc., 1976.
Morrison, Kenneth. *Reader's Work: Devices for Achieving Pedagogic Events in Textual Materials for Readers as Novices to Sociology.* Doctoral Dissertation, Department of Sociology, York University, 1976.
Morrison, Kenneth. 'Some Researchable Recurrences in Sciences and Social Science Inquiry.' Unpublished paper. Toronto: York University, 1980.
Pack, Christopher. 'Features of Signs Encountered in Designing a

BIBLIOGRAPHY

Notational System for Transcribing Lectures.' Unpublished paper, Department of Sociology, UCLA, 1975.

Pack, Christopher. 'Towards a Phenomenology of Transcription.' Unpublished paper, Department of Sociology, UCLA, 1975.

Pack, Christopher. *Interactional Synchrony Reassessed - An Attempt to Apply the Discovery of Interactional Synchrony in Neonates as an Index of Attachment in Non-organic Failure to Thrive Syndrome.* Doctoral Dissertation, Department of Sociology, UCLA, 1982.

Pollner, Melvin. 'Notes on Self-Explicating Settings.' Unpublished paper, Department of Sociology, UCLA, 1970.

Quine, Willard Van Orman. *Mathematical Logic*, revised edn. Cambridge: Harvard University Press, 1951.

Robbin, Joel W. *Mathematical Logic: A First Course.* New York: Benjamin/Cummings Publishing Company, 1969.

Robillard, Albert R. 'Applied Behavioral Analysis.' Unpublished paper, Department of Human Development, Michigan State University, 1980.

Robillard, Albert R. and Pack, Christopher. 'The Clinical Encounter: The Organization of Doctor-Patient Interaction.' Unpublished paper, Department of Human Development, Michigan State University, 1980.

Robillard, Albert R., White, Geoffrey M. and Maretzki, Thomas W. 'Between Doctor and Patient: Informed Consent in Conversational Interaction.' Unpublished paper, University of Hawaii, Department of Psychiatry, 1981.

Rosser, Barkley. 'Extensions of Some Theorems of Gödel and Church.' *Journal of Symbolic Logic* 1, 3 (1936): 87-91.

Rosser, Barkley. 'An Informal Exposition of Proofs of Gödel's Theorems and Church's Theorem.' *Journal of Symbolic Logic* 4 (1939): 53-60.

Sacks, Harvey. 'On the Analyzability of Stories by Children.' In John J. Gumperz and Dell Hymes (eds), *Directions in Sociolinguistics: The Ethnography of Communication.* New York: Holt, Rinehart and Winston, 1972.

Sacks, Harvey. 'Unpublished Lectures.' Department of Sociology, UCLA and School of Social Sciences, University of California, Irvine, 1964-75.

Sacks, Harvey, Schegloff, Emanuel and Jefferson, Gail. 'A Simplest Systematics for the Organization of Turn-Taking on Conversation.' *Language* 50 (1974): 696-735.

Sah, Chih-Han. *Abstract Algebra.* New York: Academic Press, 1967.

Schegloff, Emanuel. 'Sequencing in Conversational Openings.' *American Anthropologist* 70 (1968): 1075-95.

Schegloff, Emanuel. 'Notes on a Conversational Practice: Formulating Place.' In David Sudnow (ed.), *Studies in Social Interaction.*

Schegloff, Emanuel and Sacks, Harvey. 'Opening Up Closings.' *Semiotica* 8 (1973): 289-327.

Schenkein, Jim (ed.), *Studies in the Organization of Conversational Interaction.* New York: Academic Press, 1978.

Schrecker, Friedrich. 'Doing a Chemical Experiment: The Practices of Chemistry Students in a Student Laboratory in Quantitative Analysis.' Unpublished paper, Department of Sociology, UCLA, 1980. To appear in Harold Garfinkel (ed.), *Studies of Work in the Discovering Sciences*.

Shoenfield, Joseph R. *Mathematical Logic*. Reading: Addison-Wesley Publishing Company, 1967.

Skolem, Thoralf. 'Begründung der elemetaren Arithmetik durch die rekurrierende Denkweise ohne Anwendung scheinbarer Veränderlichen mit unendlichem Ausdehnungsbereich.' (The Foundations of Elementary Arithmetic Established by Means of the Recursive Mode of Thought, Without the Use of Apparent Variables Ranging over Infinite Domains.') *Videnskapsselskapets skrifter, I. Matematisknaturvidenskabelig klasse*, no. 6 (1923). English translation appearing in Jean van Heijenoort. *From Frege to Gödel: A Source Book on Mathematical Logic, 1879-1931*, pp. 303-33.

Spivak, Michael. *Calculus on Manifolds*. Menlo Park: W.A. Benjamin, Inc., 1965.

Sudnow, David (ed.), *Studies in Social Interaction*. New York: The Free Press, 1972.

Sudnow, David. *Ways of the Hand*. Cambridge: Harvard University Press, 1978.

Sudnow, David. *Between Two Keyboards*. New York: Alfred Knopf and Company, 1980.

Szabo, M.E. (ed.), *The Collected Papers of Gerhard Gentzen*. Amsterdam: North-Holland Publishing Company, 1969.

Takeuti, Gaisi, *Proof Theory*. Amsterdam: North-Holland Publishing Company, 1975.

Tarski, Alfred. 'The Concept of Truth in Formalized Languages.' Translated from the German translation of the original Polish text (1933) by J.H. Woodger in A. Tarski, *Logic, Semantics, and Metamathematics: Papers from 1923 to 1938*, pp. 152-278.

Tarski, Alfred. *Logic, Semantics, and Metamathematics: Papers from 1923 to 1938*. Trans. J.H. Woodger. Oxford: Clarendon Press, 1956.

Turner, Roy (ed.), *Ethnomethodology, Selected Readings*. Baltimore: Penguin Books Inc., 1974.

van Heijenoort, Jean. *From Frege to Gödel: A Source Book in Mathematical Logic, 1879-1931*. Cambridge: Harvard University Press, 1967.

Whitehead, Alfred North and Russell, Bertrand. *Principia Mathematica*. Cambridge: Cambridge University Press. Vol. I, 1910 (second edn, 1925). Vol. II, 1912 (second edn, 1927). Vol. III, 1913 (second edn, 1927).

Wieder, D. Lawrence. 'Behavioristic Operationalism and the Life-World: Chimpanzees and Chimpanzee Researches in Face-to-Face Interaction.' *Sociological Inquiry* 50 (1980): 75-103.

Wilder, Raymond. *The Foundations of Mathematics*. Second edn, New York: John Wiley and Sons, 1965.

50,647

```
QA        Livingston, Eric.
8.4
.L58      The
1985        ethnomethodological
            foundations of
            mathematics
```

MCD
APR 13 2000
MAR 30 2000

CAMROSE LUTHERAN COLLEGE
LIBRARY

© THE BAKER & TAYLOR CO.